Lipids in Health and Nutrition

Lipids in Health and Nutrition

Edited by

J.H.P. Tyman
Brunel University, Uxbridge, UK

Based on the proceedings of a two-day symposium of the Lipid Group of the RSC Perkin Division on Lipids in Health and Nutrition held on 9-10 September 1996 at Sheffield Hallam University, UK

Special Publication No. 244

ISBN 0-85404-798-0

A catalogue record of this book is available from the British Library

© The Royal Society of Chemistry 1999

All rights reserved
Apart from any fair dealing for the purposes of research or private study, or criticism or review as permitted under the terms of the UK Copyright, Designs and Patents Act, 1988, this publication may not be reproduced, stored or transmitted, in any form or by any means, without the prior permission in writing of The Royal Society of Chemistry, or in the case of reprographic reproduction only in accordance with the terms of the licences issued by the Copyright Licensing Agency in the UK, or in accordance with the terms of the licences issued by the appropriate Reproduction Rights Organization outside the UK. Enquiries concerning reproduction outside the terms stated here should be sent to The Royal Society of Chemistry at the address printed on this page.

Published by The Royal Society of Chemistry,
Thomas Graham House, Science Park, Milton Road, Cambridge CB4 0WF, UK

For further information see our web site at www.rsc.org
Typeset by Computape (Pickering) Ltd, Pickering, North Yorkshire, UK
Printed by MPG Books Ltd, Bodmin, Cornwall, UK

Preface

This book is broadly based on a two-day meeting of the Lipid Group of the Perkin Division of the Royal Society of Chemistry entitled 'Lipids in Health and Nutrition' which was held at Sheffield Hallam University in September 1996.

The lecturers were invited to write formal accounts of their contributions and at that time the majority agreed. However, since then, nearly one half found themselves unable for one reason or another to fulfil their original intentions and rather than allow the very interesting proceedings to sink into obscurity the editor was obliged to invite others to participate and keep the topic afloat. The result has been the present account.

Although nutrition and health are truly two enormous areas of interest and the role of lipids in those subjects might appear to be one of many equally important subdivisions, the consequences of 'fat' for the well being of living creatures has assumed what may seem to be a disproportionate amount of attention. Newspaper articles abound and expatiate on it, television series such as 'Fat' (ITV) have proved penetrating and educational; even at the 1998 Christmas Royal Institution lectures for children by Nancy Rothwell, a renowned researcher on adiposity gave her young audience an account of aspects of diet as well as many other fascinating excursions into biology. The symposium on 'Lipids and Nutrition' by the Oils and Fats group of the Society of Chemical Industry in 1996 attracted much interest, as did the Conference in 1994 at the University of Ulster on 'Nutrition and Sport' at which the role of lipids was one of the many sections. Several books have appeared recently, including 'Food Lipids and Health' (eds. R. E. McDonald and D. B. Min) and 'Fatty Acids in Food and their Health Implications', by Ching Kuang Chow, both published by Dekker.

The title of the present book involving lipids in both nutrition and health must desirably be multidisciplinary while many aspects of lipid interests are chemical and technological. Here the problems begin. A middle ground is opened up implicating biochemistry, biology, pharmacy and medicine, an area which is not necessarily a no-mans-land but one where the chemist treads with caution. The popular dietary approach to 'fats' has been somewhat obsessional and with an overemphasis on their exclusion. Thus a reaction has occurred resulting in, for example, a column by Nigel Hawkes ('The Times', December

24th, 1997) headed 'Fatty diet cuts stroke risk, say US doctors', based on an academic study by M. Gillman et. al., published in *J. Am. Med. Assoc.*, and another quoting work by R. M. Krauss in the *Am. J. Clin. Nutr.*, 1997, **65**, 885 that the response to low-fat diets depends on the individual's genes and their value should not be taken as an article of faith. Findings by P. Diehr et al., *Am. J. Public Health*, 1998, **88**, 623–629, that older 'fatties' live longer according to an American survey of the over 65s and summarised in 'The Times' for April 10th, 1998, suggest that the issues are not straightforward and almost certainly have previously been oversimplified, resulting often in arbitrary and draconian advice and measures.

The health/disease aspect of dietary fats has tended to be submerged in a preoccupation with age/weight, while interest in the biosynthesis of lipidic materials in the body has received less attention. Nevertheless, the recent volume of media interest and of academic published work has not only concentrated interest in the role of lipids in health and nutrition, but tended to inform that lipids of all sorts have a very wide occurrence and variety of functions in the human body.

No small book could expect to detail all these complex aspects and in the present volume the choice of subject matter is selective, yet each author has contributed an extensive bibliography, often covering associated areas of the subject. The published results of scientific and statistical work may sometimes appear to be contradictory, which serves to show that the truth is far from easy to unravel and that trends only can be indicated.

During the past four or more decades the production of lipids in the form of triacylglycerols has almost quadrupled following the increase of world population, but their usage has increased most significantly in the West. The perceived importance of polyunsaturated fatty acids has led both to an increased production of the corresponding conventional triacylglycerols and to new crops containing higher amounts of desired components.

The whole area has been reviewed by Frank Gunstone in an introductory chapter, 'Lipids: Global Resources and Consumption', giving an overview of the current resource situation. In 'Studying Lipid Metabolism using Stable Isotopes', Charles Scrimgeour describes advanced analytical techniques for following the fate of lipids in the living system. In a previous volume in this series, elegant synthetic methods had been described by L. Crombie. The potential of stable isotopes incorporated as tracers in compounds for lipid metabolic studies has been expanded by recent developments in commercially available gas chromatography-isotope ratio mass spectrometric (GC-IRMS) systems, allowing accurate measurement of very low levels (ca. 0.01%) of ^{13}C tracer in ng amounts of individual fatty acids. Recent developments involving deuterated tracers and site specific ^{13}C analysis of fatty acids are described and applications discussed. Health and disease effects of *trans* polyunsaturated fatty acids are reviewed by David Kritchevsky in his contribution. *Trans* isomers of polyunsaturated fatty acids are generally produced by heat or frying treatments and in deodorisation of oils and to a lesser extent by catalytic hydrogenation. For *Z,Z,Z*-(9:12:15)-linolenic acid the major isomers formed

Preface

during deodorisation are the $9E$ and the $15E$ while Z,Z-(18:2)-linoleic acid gives $12E$ and the $9E$ isomers in addition to some $9E,12E$. The 18:2-$9Z,12E$ isomer is metabolised to the 20:4-$5Z,8Z,11Z,14E$ isomer of arachidonic acid and the $15E$ isomer of linolenic acid is elongated to the $17E$ isomer of EPA and to the $19E$ isomer of DHA. All these *trans* polyunsaturated acids are capable of potential incorporation into tissue lipids. The physiological significance of stereochemical modifications in the healthy and diseased human system is described.

The role of antioxidants in promoting health is universally applauded, yet in dietary regimes recommending a much reduced fatty diet the intake of the lipidic antioxidant vitamins is curtailed, resulting in potential avitaminosis unless accompanied by separate vitamin supplementation. The importance of other antioxidant sources such as the flavonoids has long been recognised and promoted in many dietary and beverage recommendations advocating fruit, vegetables, tea and wine. Michael Gordon and Andrea Roedig-Penman have studied the role of a number of separate flavonoids in oil and emulsion systems and the effects of heat, metal ions, chelating agents and of the presence of tocopherols on certain members such as quercetin. Although the use of a cocktail of flavonoids is likely to occur in the diet and appears desirable, the scientific study of individual compounds under a variety of conditions is a vital link to establish individual properties.

The chapter by Christine Williams is concerned with the influence of dietary fatty acids in coronary heart disease. Public health advice on diet for prevention of coronary heart disease, and therapeutic diets for the treatment of patients with it, have previously been based on the cholesterol hypothesis which links effects of dietary fatty acids to the pathogenesis of atherosclerosis through the influence of fat and different fatty acids on circulating cholesterol levels. As a result, in the 1980s the consumption of low fat diets with higher polyunsaturated to saturated fatty acid ratios was recommended, based on the reported ability of some fatty acids to raise and of others to lower circulating cholesterol concentrations. It is believed now that the effects of dietary fatty acids on circulating triglyceride-containing lipoproteins are important and that the nature of the fatty acids in a meal and the background diet can profoundly influence the extent and duration of postprandial lipaemia. However, the effects of other factors such as physical activity, meal fat content and meal frequency and the type of carbohydrate in the diet which attenuate postprandial triglyceride concentrations also deserve attention.

There is little doubt that obesity has increased throughout the world, notably in America, during the last two decades and its contributory influence towards heart disease, high blood pressure, strokes and some forms of cancer is now acknowledged. Obesity is regarded as the second most important preventable cause of death in the developed world, with the result that the media in both TV and the press abound in programmes and articles on its origin and cure. Subsidiary causes other than genetic make-up such as a wholly sedentary lifestyle, fast food eating and food picking are implicated and well-documented with recommendations for altered regimes. In his article 'Lipids and Obesity', John Tyman gives an account of chemical approaches to both antiatherosclerotic drugs for cholesterol-lowering and in the control of obesity. There are numerous aspects under consideration on this complex topic and the outcome of the relative efficacy of self-regulation for the average obese person versus medication for the severely obese in an affluent and indulgent society remains debatable. As with many facets of modern life where the imparting of technical information can be regarded as either educational or completely superfluous, the former is chosen to reveal the considerable extent of pharmaceutical activity in this subject area.

'Docosahexaenoic acid (DHA): a Dietary Factor Essential for individuals with Dyslexia, Attention Deficit Disorder and Dyspraxia?' is an article by Jacqueline Stordy, in which the results of a scientific study in these controversial regions are discussed. Between 5 and 10% of the population suffer from dyslexia, attention deficit disorder or dyspraxia and children and adults with these conditions have great difficulty concentrating, have poor short-term memories and, for some, reading and spelling are very difficult. Until recently, multisensory teaching has been the most popular remedy for dyslexia and Ritalin, an amphetamine derivative, for attention deficit disorder. Research has now shown that the metabolism of fatty acids may be defective in these conditions and these aspects are discussed by the author of this chapter. It is of interest, although unconnected with this article, that a related material, eicosapentaenoic acid (EPA), has been found to cause regression of cachexia— a condition of severe weight loss common to all advanced cancer patients, a finding stemming from earlier work (*Nature*, 1996, **379**, 739–742) by M. Tisdale and his group.

The role of lipidic materials in combatting disease is legendary as, for example, with chaulmoogra oil, which was employed for centuries in China and India for the treatment of leprosy and tuberculosis, both of which result from fat-encapsulated acid-fast bacteria. It was an early finding of R. Adams from a study of the synthesis of a range of cyclopentano and branched-chain fatty acids that *in vitro* leprocidal antibacterial activity probably resulted from impairment of the fatty envelope of the bacilli.

The therapeutic importance of the prostaglandins and prostacyclins, a group of compounds which has tended to be neglected in recent years, is described by Derek Clissold, who summarises recent advances relevant to the pharmaceutical and biotechnology industries. His review concentrates on the increasing number of medical applications, the products under development and entering the market in this active area misconceived by many as moribund.

'The Importance of Mycobacterial Lipids' is a contribution by David Minnikin, a pioneer in the study of the complex lipidic components of the cell envelope of mycobacteria. Tuberculosis is a currently increasing disease throughout the world and both pharmaceutical and biochemical approaches to its combatting are active investigations. Early work by R. Anderson, sponsored by the American Tuberculosis Association, and also J. Cason established the presence of tuberculostearic, phthioic and phthienoic acids, while N. Polgar and co-workers independently established the same findings.

Preface

Greater complexity has been revealed more recently involving structures such as phthiocerol dimycocerosates, phenolic glycolipids, mycolates and other associated features, all of which are discussed by the author in a comprehensive and authoritative account of the topic.

The editor would have welcomed more contributions from specialists in other areas of lipid chemistry and biochemistry, but unfortunately many could not devote their time to this task.

He thanks all those who have contributed and given their time freely to make results available for general perusal, doubtless also open to critiscism yet nevertheless affording matters for continued discussion. I also thank the Royal Society of Chemistry for their processing and production of this book, Clare Lucas of RSC Search services for certain Chemscan profiles and Dr David Horrobin, now of Laxdale Ltd., for valuable information.

Previous titles in this series are 'Surfactants in Lipid Chemistry', 'Developments in the Analysis of Lipids' and 'Synthesis in Lipid Chemistry'.

J. H. P. Tyman
London

Contents

1	Lipids: Global Resources and Consumption *F. D. Gunstone*	1
2	Studying Lipid Metabolism Using Stable Isotopes *C. M. Scrimgeour*	15
3	*Trans* Unsaturated Fat in Health and Disease *D. Kritchevsky*	32
4	Antioxidant Properties of Flavonols *M. H. Gordon and A. Roedig-Penman*	47
5	Dietary Fatty Acids, Postprandial Lipaemia and Coronary Heart Disease *C. M. Williams*	65
6	Lipids and Obesity *J. H. P. Tyman*	76
7	Docosahexaenoic Acid: a Dietary Factor Essential for Individuals with Dyslexia, Attention Deficit Disorder and Dyspraxia? *B. J. Stordy*	102
8	The Potential for Prostaglandin Pharmaceuticals *D. Clissold*	115
9	The Importance of Mycobacterial Lipids *D. E. Minnikin, M. R. Barer, A. M. Gernaey, N. J. Garton, J. R. L. Colvine, J. D. Douglas and A. M. S. Ahmed*	130
	Subject Index	152

1
Lipids: Global Resources and Consumption

Frank D. Gunstone

SCOTTISH CROP RESEARCH INSTITUTE, INVERGOWRIE,
DUNDEE DD2 5DA, SCOTLAND, UK

1 Introduction

Many books devoted to lipid science start with a review of the structures of fatty acids and lipids. It could be argued that there should also be an introductory section on lipid (oil and fat) supplies. This chapter is an attempt to meet that need.

The annual production of oils and fats now exceeds 100 million tonnes of which about 79% are of vegetable origin and 21% of animal origin. The major use of this material is for food (hence its importance for health and nutrition). About 81% is consumed by humans and a further 5% is eaten by animals to produce yet more human food. The remaining 14%, representing about 14 million tonnes, is used by the oleochemical industry to produce mainly soaps and other surface active materials (90%). Other oleochemical outputs serve several purposes such as inks and surface coatings, lubricants, biodiesel, and production of epoxides and dibasic acids.

What is the origin of these lipid materials? Where are they produced? How do they differ from each other? Where are they consumed? Who has too much and who too little? What about oils other than the major sources? What changes have occurred in the sources of supply over the years and what further changes are likely in the future? In what ways are the materials supplied by nature less than optimum for their human use and how can they be improved? We will try to answer these questions.

2 Seventeen Major Oils

A company in Hamburg (ISTA Mielke GmbH) has been circulating data on production, imports and exports, disappearance, and prices of oils and fats for over 40 years and many of the figures quoted here are taken from their publications especially from Oil World Annual 1998.[1] They confine their reports to 17 major oils and exclude others, some of which are now significant on a world trade basis such as cocoa butter, rice bran oil, and tall oil and others which are emerging such as oils containing γ-linolenic acid.

The 17 major oils can be divided into four categories. There is a group of four animal fats (butter, lard, tallow from cows and sheep, and fish oils) and three groups of vegetable oils:

- *Tree crops* (coconut, oil palm products, and olive). These have to be planted and take some years to mature. Thereafter they continue to bear a harvest for many years. Consequently volumes of supply cannot be changed quickly from season to season.
- *Crops in which the oil is a minor product*. These include soya grown mainly for the protein meal which is generally, but not always, the more valuable commodity, cotton grown for its fibre, and corn grown for the cereal.
- *Annual crops grown for the oilseed itself* such as rape, sunflower, and groundnut. The supply of these commodities will depend on decisions taken each year whether to grow these or some other crop (grains). This decision is based on the planter's view of the relative profitability of the crops both in the short term (price, profit, subsidy) and the long term (agricultural productivity over years).

3 World Supplies of Major Oils—Amounts and Geographical Sources

The availability of these materials has changed over time. In developing this point it is useful to consider three groups: animal fats, vegetable oils group A, and vegetable oils group B. The terms A and B will be explained later. The following comments are based on production and forecasts covering the period 1965–2005. In this 40 year period total production is expected to rise from 32 to 115 million tonnes (a 3.6-fold increase). Animal fats should rise from 13 to 25 million tonnes (a 1.9-fold increase and a change from 42 to 22% of total production), vegetable fats group A rise from 8 to 16 million tonnes (a 2-fold increase and a change from 26 to 14% of total production), while vegetable oils group B rise from 10 to 74 million tonnes (a 7.4-fold increase and a change from 32 to 64% of total production). Though all commodities have increased in production level, only vegetable oils group B have increased their market share, and that markedly: approximately from one third to two thirds. These changes are even more marked when considering world trade (*i.e.* imports and exports) rather than total production. So what are the market leaders, the vegetable oils of group B? They are soya, the two products of the oil palm (palm oil and palmkernel oil), rape (canola), and sunflower. The remaining vegetable oils are in group A.

The annual production figures of the major oils and fats for a five year period are given in Table 1. This five year range of figures avoids fluctuations arising from good and bad harvests in any particular year. Since the annual production is now around 100 million tonnes these figures also represent the % of total production.

Table 1 Annual world production and yield of major oil and fat sources

	93/94[a]	94/95[a]	95/96[a]	96/97[a]	97/98[a]	Yield[b]
Soybean	18.3	19.8	20.3	20.8	22.4	0.41
Palm	13.8	15.1	16.1	17.3	17.3	3.60
Rape	9.7	10.6	11.7	11.5	12.0	0.56
Sunflower	7.5	8.4	9.2	9.3	8.9	0.51
Groundnut	4.2	4.4	4.3	4.3	4.0	0.42
Cottonseed	3.5	3.8	4.1	4.0	4.0	0.19
Olive	1.9	2.0	1.6	2.8	2.6	
Corn	1.7	1.8	1.8	1.9	1.9	
Sesame	0.7	0.7	0.8	0.8	0.8	
Linseed	0.6	0.7	0.7	0.7	0.7	
Coconut	2.9	3.5	3.0	3.3	3.3	0.36
Palmkernel	1.8	1.9	2.0	2.1	2.2	0.48
Castor	0.4	0.5	0.5	0.5	0.5	
Butter	5.7	5.7	5.7	5.7	5.8	
Lard	5.6	5.8	6.0	6.1	6.4	
Tallows	7.4	7.5	7.4	7.4	7.5	
Fish	1.5	1.4	1.4	1.3	0.9	
Total	88	92	96	100	102	

Source: Oil World Annual 1998.[1]
[a] Related to the harvest year October to September.
[b] Tonnes/hectare.

It is of interest that the leading oils are grown in different geographical locations. Soybeans are grown mainly in the USA, Brazil, Argentina, and China; palm and palmkernel in Malaysia and Indonesia; rape in China, India, the EU (15 countries), and Canada; and sunflower in the ex-USSR, Argentina, the EU (15 countries), China, India, Turkey, Eastern Europe, and the USA. Surplus supplies are usually exported but national demand is so high in the populous countries of China and India that they are net importers despite high local production. The wide geographical spread helps to overcome fluctuations arising from adverse weather conditions in some parts of the world.

4 Productivity

The final column in Table 1 shows the yield of lipid per unit area of production. These are average figures and cover a wide range of values which vary with the condition of the soil, climate, agricultural input, agricultural skills, etc. The point has already been made that some of these plants produce a second valuable commodity. Despite these caveats it is clear that the oil palm, with its two products at 4.1 tonnes per hectare, is much more productive than other lipid-producing crops at 0.2–0.6 tonnes per hectare. This fact underlines the importance of this crop and its significance in feeding the growing populations of the world. Also important in this connection is the high proportion of palm oil that is *exported* (in the range 66–75% of production during the last five years) compared with lower proportion of soybean oil (24–32% in the last five years). Put another way: although production of soybean oil at 18–22 million tonnes in the last five years exceeds that of palm oil (14–17 million tonnes) the order is reversed for exports (4–7 for soybean oil and 10–12 for palm oil).[1]

5 Consumption (Disappearance)

Before discussing the consumption of oils and fats it is useful to ask whether there is any information about what the human body actually needs. The FAO has indicated a minimum annual requirement of 12 kg per person though this figure will vary with body weight and lifestyle. Scrimshaw[2] has suggested that minimum requirements for protein, carbohydrate, and fat are 8, 10, and 15% of dietary energy intake respectively. In practice the values must be greater than this since they must total 100 but individually they should not fall below these values.

When considering *average* consumption values it must be appreciated what these signify. Most of the population will have a consumption lying between one half and twice the average value so that even with a satisfactory average value some will be getting too little and some will be getting too much. A small number of persons will even have consumption levels outside this range.

Some countries, from time to time, have carried out surveys to discover the pattern of fat consumption. These generally describe consumption for different groups, for example male and female adults, and distinguish between various fat sources such as vegetable oils consumed as such, meat, dairy products, etc.[3]

However, Oil World publications[1] provide useful information in terms of 'disappearance' on a countrywide basis (Table 2). They use this term to describe usage of all fat on a per person basis:

Disappearance = [production + imports − exports] ÷ population

It thus represents, on a personal basis, the total amount of fat used in a country. It *excludes* fat from sources outside the 17 major oil sources, but it *includes* all waste (whether by trade or by household) and also the oils and fats used by the oleochemical industry. This last represents 20% of all supplies and is not uniformly distributed. Oleochemical industries, previously concentrated in North America, Western Europe, and Japan, are now developing strongly in South East Asia. With these provisos the figures are illuminating and they are used as the basis of the following comments.

Annual consumption per head of population is rising steadily on a global basis: from 15.5 kg/person (1993) to 17.3 kg/person (1998). Rising oil and fat production is thus keeping pace with rising population. However, these average figures hide large variations between and within individual countries.

The highest figures (44–48 kg per person) relate to the USA and to Western Europe. In respect of dietary intake these figures will have to be reduced by more than 20% to allow for oleochemical production. The figures for Australia and New Zealand (29–32 kg) are probably more representative of advanced

Table 2 *Annual disappearance (kg/person) and population (millions) in selected countries*

	93	94	95	96	97	98	Pop.
World	15.5	15.7	16.2	16.6	17.0	17.3	5930
USA	44.2	45.3	43.9	44.8	45.6	47.6	274
EU (15 countries)	40.3	40.6	41.8	42.7	43.6	44.3	374
Australia	30.0	29.7	31.0	30.8	31.7	32.1	18
New Zealand	29.5	29.1	29.0	28.1	28.8	28.7	4
Japan	20.5	20.4	20.7	20.8	21.5	21.6	126
India	7.9	8.1	8.6	9.1	9.6	9.7	976
China	8.3	9.2	10.4	11.2	11.9	12.4	1233
ex USSR	17.3	14.8	13.7	14.0	13.0	13.3	294
Sudan	6.7	6.7	7.2	7.0	6.8	6.8	29
Bangladesh	4.7	4.8	5.0	4.9	5.0	5.0	124

Sourc: Oil World Annual 1998[1] which includes data for 175 countries.

countries with little oleochemical production. Other figures indicate that average daily consumption of fat in the UK is 102 g for a man (~75 kg weight) and 74 g for a woman (~ 60 kg weight). These correspond to annual intakes of 37 and 27 kg respectively.

The Japanese figure is also raised by reason of the strong oleochemical industry in that country. This is an advanced country with dietary habits different from those in the West. Nevertheless average consumption has risen considerably since World War II.

Figures for the countries of the former USSR have declined and continue to do so, reflecting the economic difficulties of the region.

The highly populated countries of India and China still have very low *average* levels of consumption. If these were increased to 20 kg per person then India would require a further 10.1 million tonnes and China a further 9.4 million tonnes. The additional production of 19.5 million tonnes would take some years to achieve.

At the low end of the scale are poor countries like Sudan (~7 kg) and Bangladesh (~5 kg) which are well below the level of 12 kg recommended by the FAO. Those who live under these conditions are not interested in the niceties of saturated, monounsaturated, polyunsaturated acids, *trans* isomers, or $n-6/n-3$ ratios. Their need is for calories and this is best met by fat whatever its detailed nature.

6 Fatty Acid Composition

The use to which lipids are put depends on their physical, chemical, and nutritional properties which depend in turn on their fatty acid and triacylglycerol composition. The latter is important but can be quite complex and for most practical purposes lipids are discussed in terms of their fatty acid composition. The availability of the major oils and fats cannot, therefore, be separated from their fatty acid composition and typical values for a range of

Table 3a *Typical fatty acid composition (%wt)*

	14:0	16:0	16:1	18:0	18:1	18:2	18:3
Cocoa butter	–	26	–	34	35	–	–
Corn	–	13	–	3	31	52	1
Cottonseed	–	27	–	2	18	51	tr[c]
Groundnut	–	13	–	3	38	41	tr
Linseed	–	6	–	3	17	14	60
Olive	–	10	–	2	78	7	1
Palm	–	46	–	4	40	10	tr
Palm olein	–	40	–	4	43	11	tr
Rape (high erucic)[a]	–	3	–	1	16	14	10
Rape (low erucic)	–	4	–	2	56	26	10
Soybean	–	11	–	4	22	53	8
Sunflower	–	6	–	5	20	60	tr
Sunola	–	4	–	5	81	8	tr
NuSun	–	4	–	5	65	26	–
Butter[b]	12	26	3	11	28	2	1
Lard	2	27	4	44	11	–	–
Beef tallow	3	27	11	7	48	2	–
Mutton tallow	6	27	2	32	31	2	–

[a] Also 20:1 6% and 22:1 5%.
[b] Also 4:0 3%, 6:0 2%, 8:0 1%, 10:0 3%, 12:0 4%.
tr = trace (<1%).

Table 3b *Typical fatty acid composition (%wt) of lauric oils*

	8:0	10:0	12:0	14:0	16:0	18:0	18:1	18:2
Coconut	8	7	48	16	9	2	7	2
Palmkernel	3	4	45	18	9	3	15	2
Lauric canola[a]	-	-	39	4	3	2	33	11

[a] Also 18:3 6%.

oils and fats are collected in Table 3a. These will not be considered in detail but a few general points are made. Cited figures taken from Tables 3a and 3b must be considered merely as typical values.

Coconut and palmkernel oils (Table 3b) are typical lauric oils which differ from most other vegetable oils. They are important both in the food industry and the oleochemical industry and are characterised by high levels of lauric acid (12:0) with significant levels of myristic acid (14:0) and useful quantities of the shorter C_8 and C_{10} acids. These oils are very rich in saturated acids and contain very little unsaturated acid. Palmkernel oil is one of two products of the oil palm and must not be confused with the very different palm oil which is the major product of this tree. Attempts are being made to domesticate some Cuphea oils which are also rich in short and medium chain acids and a rape, genetically modified to produce a lauric oil, is already being grown and harvested in USA.

Most vegetable oils contain mainly palmitic, oleic, and linoleic acids (Table 3a). Palmitic acid is the major saturated acid and reaches significant levels in palm oil (40%) and cottonseed oil (27%). Some oils are rich in oleic acid (olive, canola), some in linoleic acid (corn, cottonseed, soybean, sunflower), and some in both of these acids (groundnut). Seed breeders have produced oleic-rich varieties as with sunflower which is now available with 80% oleic acid (high oleic sunflower, sunola) and with 65% oleic acid (NuSun). This latter variety is expected to become a major source of sunflower oil in the USA.[4]

Linolenic acid is the major component of linseed oil (60%) and is the basis of most of the industrial uses of this oil. It is also present in soybean oil (~8%) and rapeseed oil (~10%). There is some ambivalence toward this acid. As it is responsible for many of the undesirable aromas and flavours that develop in linolenic-containing oils there is a desire to reduce its level, but with growing awareness of a dietary need for $n-3$ acids such as linolenic, eicosapentaenoic, and docosahexaenoic there is an opinion that these needs are met better by linolenic-containing vegetable oils than by fish oils rich in longer $n-3$ polyunsaturated acids such as eicosapentaenoic and docosahexaenoic acids.

The major animal fats are more saturated than vegetable oils and contain very little polyunsaturated acids. Saturated acids are generally 40–60% and monounsaturated acids 30–60%. Butter has acids with a wide range of chain length (4–18 carbon atoms) but like the animal depot fats it is rich in saturated and monoene acids and low in polyunsaturates.

Cocoa butter, the lipid component in chocolate, is an unusual vegetable fat with saturated (~60%) and monoene (~35%) acids in such proportion that its triacylglycerols are mainly of the type SOS (S = saturated, O = oleic) and are responsible for the characteristic melting behaviour of this fat which is so important in chocolate.

These comments hold for the 17 major oils and fats and also for many of the less common seed oils. However, some oilseeds illustrate the rich diversity of plants in generating other unusual fatty acids, sometimes at very high levels. Examples include castor oil with 90% of ricinoleic acid (12-hydroxy-9-octadecenoic acid), coriander oil with 80% of petroselinic acid (6-octadecenoic acid), *Vernonia galamensis* seed oil with ~75% of vernolic acid (12,13-epoxy-9-octadecenoic acid), and the seed oil of *Picramnia sow* with 95% of tariric acid (6-octadecynoic acid).

Fish oils are characterised by the wide range of acids present and particularly by the highly unsaturated members. Saturated (14:0 and 16:0), monenoic (16:1, 18:1, 20:1, and 22:1), and $n-3$ polyenoic acids (eicosapentaenoic, 20:5, and docosahexaenoic, 22:6) are frequently major components.

7 Extending the Use and Supply of Natural Lipids

For the most part the lipids that are consumed and which are part of the health and disease issue are not quite the native materials of vegetable or animal origin.

Extraction and Refining

Lipid must first be extracted and then usually refined by a series of processes which may include degumming, neutralisation, bleaching, deodorisation, and physical refining. Through these processes, undesirable constituents are removed along with some desirable constituents. Palatability is improved but some nutritive value is lost. However, phospholipids, tocopherols, and phytosterols, present in by-product streams, are recovered and used.

Even then, the refined products are not always wholly satisfactory for their end use and for at least a century oils and fats have been processed in several ways to produce more useful products. The parameters which have to be met are mainly nutritional (getting the 'correct' blend of fatty acids in appropriate triacylglycerol form) and physical (getting the appropriate melting profile and crystalline habit to make good quality spreads). Attempts to 'improve' what nature supplies are either technological or biological. The former include blending, fractionation, hydrogenation, and interesterification; the latter include domestication of wild crops, conventional seed breeding, genetic modification, and the use of microbial sources of lipids.[5]

Technological Processes

The technological processes of blending, fractionation,[6] hydrogenation,[7] and interesterification with chemical or enzyme catalysts[8] will not be described here. It should be noted that partial hydrogenation produces acids with *trans* unsaturation. These are considered undesirable from a dietary point of view and fractionation and interesterification are being used as alternative procedures to provide products with little or no *trans* acids but with slightly elevated levels of saturated acids.[9] On the other hand, hydrogenation processes are continually being improved to reduce the level of *trans* acids.[10] All these processes can be carried out on a large scale and most spreads and baking fats have been modified in one or other of these ways.

Biological Processes

The biological techniques are both old and new. Seed breeding has been used to develop plants which are easier to grow and harvest and which have higher yields and more appropriate composition. Examples include the production of double zero rapeseed oil (canola oil, low in erucic acid and in glucosinolates), linola (high linoleic acid oil), and high-oleic sunflower and safflower oils.

Domestication of Wild Crops. Attempts are being made to domesticate several wild crops with emphasis on those producing seed oils very rich in a single fatty acid either of conventional type such as oleic or of a less conventional nature. These are being developed mainly for the oleochemical industry and some examples are given in Table 4. However *Camelina sativa*

Table 4 *Some new seed oils of potential interest*

Species	Major fatty acid	Content (%)
Cuphea species	8:0, 10:0, 12:0, or 14:0	High
Coriander	Petroselinic (6c-18:1)	80
Euphorbia lathyris	Oleic	84
Calendula	Calendic (8t,10t,12c-18:3)	58
Camelina	—[a]	—[a]
Meadowfoam	5–20:1	63–66
(*Limnanthes alba*)	5–22:1	2–4
	13–22:1	8–12
	5,13–22:2	16–18
Crambe	Erucic (13c-22:1)	55–60
Lesquerella	Lesquerolic (14-OH-11c-20-1	50-55
Dimorphotheca	Dimorphecolic (9-OH-10t,12c-18:2)	62
Vernonia galamensis	Vernolic (12,13-epoxy-9c-18:1)	80
Euphorbia lagascae		65

[a] See text

seed oil (also called false flax), which contains a useful quantity of α-linolenic acid, is being developed as a dietary source of this acid and the oil has received food approval in France and the UK.[11,12] In addition to its interesting fatty acid composition this plant attracts attention because it grows well with lower inputs of fertilisers and pesticides than more traditional crops like rape and linseed. The plant can also be grown on poorer soils and shows better gross margins than the other two plants after allowing for direct costs and subsidy payments. The seed yield is in the range 1.5–3.0 tonnes per hectare and the oil content between 36 and 47%. The oil has an unusual fatty acid composition. It contains significant levels of oleic acid (12–14%), linoleic acid (16–24%), linolenic acid (30–40%), and of C_{20} and C_{22} acids, especially 20:1 (15–23%). Despite its high level of unsaturation it shows reasonable oxidative stability. Attempts are being made to optimise the agronomy.[13]

Genetic Modification. The rape plant seems to lend itself to genetic manipulation and the first genetically modified oilseed with changed fatty acid profile was canola oil containing lauric acid. This was developed by Calgene and the crop is being grown in the USA, though successful field trials have been conducted elsewhere. To obtain this new oil Calgene scientists isolated the transesterase which produces lauric acid in the Californian Bay tree and transferred it to the rape plant. When introduced into rapeseed the resulting oil contained more than 50% of lauric acid though this was somewhat reduced in the commercial crop (Table 3b). To go beyond this level it is necessary to introduce a further gene (lysophosphatidic acid acyl transferase, LPAT) which will promote the acylation of the *sn*-2 position with lauric acid.

Other oils at various stages of development include the following:

- Rapeseed oils still higher in lauric acid, high in erucic, palmitic, oleic, or linoleic acid, or containing C_8 and C_{10} acids, myristic, stearic, petroselinic,

ricinoleic, vernolic, or γ-linolenic acid, and also wax esters in place of the normal triacylglycerols.
- Soybean oils with lower saturated acids, lower linolenic acid, and higher stearic acid, etc., as well as seeds producing meal of enhanced nutritional value.
- Sunflower oil with high palmitic, stearic, oleic, or linoleic acid.
- Corn oil with high oleic acid.

The level of linolenic acid is being reduced because its oxidation leads to undesirable flavours. Saturated acids are being increased to produce oils which can be used to make spreads without partial hydrogenation.

Oils from Micro-organisms. Lipids have been obtained traditionally from animal and plant sources. However some valuable lipids are now being produced from micro-organisms. These can be grown by fermentation in tanks or in ponds and harvested. Lipids are extracted from the dried mycelium. Because of the costs associated with these processes they are of commercial interest only for lipids rich in the higher polyunsaturated fatty acids (PUFA). Such oils are free of cholesterol, heavy metals, and pesticides and are generally simpler in lipid composition than the animal lipids and fish oils which serve as the alternative sources of most of these PUFA.

Micro-organisms are often rich sources of lipids and these have been studied extensively—mainly on the look–see basis—for many years. Some of these are known to be rich sources of PUFA in general and of the C_{20} and C_{22} acids in particular. There is a growing interest in these acids, especially arachidonic, eicosapentaenoic, and docosahexaenoic, following the recognition of their importance in membrane phospholipids and in the development of the brain and the visual system of the developing child, both *in utero* and in the early months and years of infancy. There is also a wider concern about the relation between the $n-6/n-3$ ratio and several diseased states.[14] There are no suitable vegetable sources of arachidonic acid, eicosapentaenoic acid, or docoasahexaenoic though these acids are present in egg lipids (arachidonic) and in fish oils (eicosapentaenoic and docosahexaenoic). In recent years microbiological sources of arachidonic acid and docosahexaenoic acid have been developed and commercial products are available for dietary purposes—at a price. These are used increasingly in Europe in infant formula, especially for pre-term babies.

Several companies are producing arachidonic acid-containing lipid as glycerol esters from the filamentous fungi *Mortierella alpina*. This is mixed with high-oleic sunflower oil to offer a standard product with 40% arachidonic acid. In one product other $n-6$ acids (14%) are also present along with saturated (15%) and monounsaturated acids (21%). The product contains no $n-3$ acids.[15]

One company is producing a docosahexaenoic acid-rich oil from the microalga *Crypthecodinium cohnii*. Typically this product contains docosahexaenoic acid (~40%), saturated acids (~36%), and monounsaturated acids (~22%). The only other polyunsaturated acid is linoleic (~1%).[16]

Eicosapentaenoic acid is present in several micro-organisms but so far no commercial product is available.

Fish oils are a rich source of eicosapentaenoic acid and docosahexaenoic acid and products are available in which the level of one or both of these acids has been enhanced by the use of techniques such as urea fractionation, molecular distillation, or selective reaction in the presence of enzyme.[17]

8 Oils Other than the 17 Major Oils

Many minor oils are available beyond the 17 that provide statistical data for Oil World. A selection of these are described as typical examples. Others are reported in a book devoted to this topic.[18]

Speciality Oils

These oils are usually available in only limited quantities and it is essential to ensure that the sources located will provide a reliable and adequate supply of good quality material. Since the oils are to be used as dietary supplements, as health foods, or as gourmet oils it is important that the seeds be handled, transported, and stored under conditions which will maintain quality and it may be necessary to consider growing crops under conditions which will minimise the level of pesticides.

Extraction can be carried out in several ways including cold-pressing at temperatures not exceeding 45 °F (7 °C), pressing at higher temperatures, and/or solvent extraction. Solvent extraction is not favoured for high quality gourmet oils. Supercritical fluid extraction is also a possibility but there is no evidence of this being used commercially for this purpose.

Some speciality oils such as walnut, virgin olive, hazelnut, pistachio, and sesame can be used as expressed, merely after filtering, but for others some level of refining is generally necessary. If the oil has a characteristic flavour of its own it may be desirable to retain this and high-temperature deodorisation must then be excluded or reduced to a minimum. Once obtained in its final form the oil must be protected from deterioration—particularly oxidation. This necessitates the use of stainless steel equipment, blanketing with nitrogen, and addition of appropriate antioxidants.[19,20] The following, among others, are available as speciality oils: almond, avocado, grapeseed, hazelnut, mustard, passionflower, pistachio, pumpkinseed, and walnut.

Some companies produce super-refined oils and phospholipids by adsorption chromatography using a batch process with several kilograms of oil in a high oil to adsorbent ratio at room temperature. This process removes most of the colour, hydroperoxides and other oxidised fragments, trace metals, and odour. Natural antioxidants may be removed at the same time but these can be added back to the refined product if desired. These highly refined oils are particularly useful in the cosmetics industry since they need less of the expensive ingredients such as colour and fragrance which are added to produce particular products. They are reported to have high oxidative stability following the removal of metallic pro-oxidants and oxidised materials. Their high quality also makes them suitable for pharmaceutical use.

Cocoa Butter

The cocoa bean is the source of cocoa powder and of a solid fat called cocoa butter, both of which are important ingredients of chocolate. The value of cocoa butter for this purpose depends mainly on its melting behaviour which is related to its triacylglycerol composition (Section 5). Because cocoa butter commands a good price which fluctuates markedly with changes in supply, usually for climatic reasons, there is an interest in the production and use of cheaper alternatives.[21] Annual production is about 2.7 million tonnes of beans containing 45–48% of fat.

Oils Containing γ-Linolenic Acid and Stearidonic Acid

Though comparatively rare, γ-linolenic acid (GLA or 6,9,12–18:3) has attracted attention because of claims concerning its beneficial effect in the treatment of several diseases.[22] There is thus a strong and growing niche market for the three oils containing GLA which are commercially available, viz. evening primrose (10% GLA), borage (starflower) (23%), and blackcurrant (17%). Evening primrose and borage grow in temperate regions of North America, Europe, and New Zealand but most evening primrose now comes from China. Production levels are not published but evening primrose oil is possibly 1200–1500 tonnes and borage oil 400–500 tonnes. These oils are subject to increasing demand and these figures may be too low. Oils with enhanced levels of γ-linolenic acid are available through enzymic enhancement.

Rice-bran Oil

Rice (*Oryza sativa*) is an important cereal with an annual production of above 500 million tonnes. To produce white rice the hull is removed and the bran layer is abraded giving 8–10% of the rice grain. This contains the testa, cross cells, aleurone cells, part of the aleurone layer, and the germ and includes almost all the oil of the rice coreopsis. There is a potential for 15 million tonnes of bran which would furnish over 3 million tonnes of rice bran oil but present production is only about 0.45 million tonnes per annum and not all this is of food grade.

Lipases liberated from the testa and the cross cells promote rapid hydrolysis of the oil and therefore it should be extracted within hours of milling.

Refined rice bran oil is an excellent salad oil and frying oil with high oxidative stability. It also finds several non-food uses. The oxidative stability of this oil is exploited in Good Fry Oil. This is a frying oil based on oleic-rich sunflower oil to which is added up to 6% of rice bran and/or sesame oil to confer high oxidative stability.

Rice bran oil is reported to lower serum cholesterol by reducing the less desirable low density lipoproteins and very low density lipoproteins without changing the level of the desirable high density lipoproteins. This effect seems not to be related to fatty acid or triacylglycerol composition but to the unsaponifiable fraction and probably to the oryzanols (1.5–2.0% of the oil). These are ferulic acid esters of sterols and triterpene alcohols.

The major acids in rice bran oil are palmitic (12–18%, typically 16%) oleic (40–50%, typically 42%), and linoleic acid (29–42%, typically 37%). The oil contains phospholipids (~5%), a wax which may be removed and finds industrial use, and unsaponifiable material including sterols, 4-methylsterols, triterpene alcohols, tocopherols, squalene, etc.[23–25]

Sesame

This is a minor oil (annual production below 1 million tonnes) grown mainly in India and China but also in Myanmar (Burma), Sudan, and Mexico. The seed contains 40–60% oil with almost equal levels of oleic (range 35–54%, average 40%) and linoleic acid (range 39–59%, average 46%) along with palmitic (8–10%) and stearic (5–6%). The oil contains sesamin which is or produces a powerful antioxidant. The high oxidative stability of this oil is responsible for its use in Good Fry oil (see above).

References

1. T. Mielke (ed), *Oil World Annual 1998*, ISTA Mielke, GmbH, Hamburg, Germany.
2. N. S. Scrimshaw, *Lipid Technol.*, 1998, **10**, 105.
3. J. Gregory, K. Foster, H. Taylor, and M. Wiseman, *Dietary and Nutritional Survey of British Adults*, HMSO, London, 1990.
4. Anon., *INFORM*, 1998, **9**, 715.
5. F. D. Gunstone, *Prog. Lipid Res.*, 1998, **37**, 277.
6. R. E. Timms, in *Lipid Technologies and Applications* (ed. F. D. Gunstone and F. B. Padley), Marcel Dekker, New York, 1997, Chapter 8, p. 199.
7. W. T. Koetsier, in *Lipid Technologies and Applications* (ed. F. D. Gunstone and F. B. Padley), Marcel Dekker, New York, 1997, Chapter 10, p. 265.
8. A. Rozendaal and A. R. Macrae, in *Lipid Technologies and Applications* (ed. F. D. Gunstone and F. B. Padley), Marcel Dekker, New York, 1997, Chapter 9, p. 223.
9. D. A. Allen, *Lipid Technol.*, 1998, **10**, 53.
10. R. C. Hastert, *Lipid Technol.*, 1998, **10**, 101.
11. A. Hebard, *Lipid Technol.*, 1998, **10**, 81.
12. C. Leonard, *INFORM*, 1998, **9**, 830.
13. J. Zubr, *Ind. Crops Products*, 1997, **6**, 113.
14. H. Okuyama, T. Kobayashi, and S. Watanabe, *Prog. Lipid Res.*, 1997, **35**, 409.
15. D. J. Kyle, *Lipid Technology*, 1997, **9**, 116.
16. D. J. Kyle, *Lipid Technology*, 1996, **8**, 107.
17. G. G. Haraldsson *et al.*, *J. Am. Oil Chem. Soc.*, 1997, **74**, 1419 and 1425.
18. B. S. Kamel and Y. Kakuda (ed), *Technological Advances in Improved and*

Alternative Sources of Lipids, Chapman and Hall, London, 1994, several chapters.
19. J. Hancock and C. Houghton, *Lipid Technol.*, 1990, **2**, 90.
20. F. D.Gunstone, *Int. Food Ingredients*, March 1994, p. 51.
21. F. B. Padley, in *Lipid Technologies and Applications* (ed. F. D. Gunstone and F. B. Padley), Marcel Dekker, New York, 1997, Chapter 15, p. 391.
22. D. F. Horrobin, *Prog. Lipid Res.*, 1992, **31**, 163.
23. B. Sayre and R. Saunders, *Lipid Technol.*, 1990, **2**, 72.
24. Anon., *J. Am. Oil Chem. Soc.*, 1989, **66**, 615.
25. Anon., *J. Am. Oil Chem. Soc.*, 1989, **66**, 620.

2
Studying Lipid Metabolism Using Stable Isotopes

Charles M. Scrimgeour

SCOTTISH CROP RESEARCH INSTITUTE, INVERGOWRIE, DUNDEE, DD2 5DA, SCOTLAND, UK

1 Introduction

Stable isotopes do not emit harmful radiation. This is commonly seen as the main reason for using stable rather than radioactive isotopes to study human metabolism. However, stable isotopes also differ from radioactive isotopes in the way they are measured, and in the quality and detail of information that can be obtained from them. In general, compound- and site-specific information is more readily obtained with stable isotopes, as compound separation and isotopic measurement are easily combined, for example by gas chromatography-mass spectrometry (GC-MS). This allows the behaviour of particular metabolites and functional groups within these compounds to be studied quantitatively without additional preparative or degradative chemistry.

The recent commercial availability of gas chromatography-combustion-isotope ratio mass spectrometry (GC-C-IRMS) is of particular relevance to studying lipid metabolism. GC-C-IRMS based techniques are capable of extremely precise measurements of very low levels of ^{13}C in GC-separated components. Compound-specific analysis of fatty acids by GC-MS or nuclear magnetic resonance (NMR) requires substantial stable isotope enrichment. GC-C-IRMS, in contrast, can follow low tracer doses into increasingly dilute metabolic pools, and into desaturated, elongated and retro-converted compounds.

2 Stable Isotopes and Stable Isotope Tracers

Isotopes are forms of an element which differ in the number of neutrons in the nucleus. The chemical properties of an element (metal, non-metal, etc.) are determined by the atomic number, and this is equal to the number of protons in the nucleus. The atomic mass depends on the number of protons and

Table 1 Isotopes of hydrogen and carbon

	Abundance (%)	Spin	Half-life	Decay
Hydrogen				
^1H	99.984	0.5		Stable
^2H	0.015	1		Stable
^3H	0	0.5	12.3 years	$\beta-$
Carbon				
^{11}C	0	1.5	20.3 min	$\beta+$
^{12}C	98.892	0		Stable
^{13}C	1.108	0.5		Stable
^{14}C	0	0	5730 years	$\beta-$
^{15}C	0	0.5	2.45 s	$\beta-$

neutrons, and increases by one for each additional neutron. Some isotopes are unstable and are radioactive, but many elements exist as mixtures of two or more stable isotopes. Additional neutrons in the nucleus leave the element's fundamental chemistry unchanged, but the increase in atomic mass may alter the kinetic or equilibrium behaviour of molecules containing different isotopes.

Lipid molecules are usually labelled with isotopes of carbon or hydrogen. In both these elements the most abundant stable isotope (~99% or more) is the one of lowest atomic mass, and the other stable isotope is one unit heavier (Table 1). This pattern of stable isotopes is not however a general rule for all elements. Tracer molecules contain a higher proportion of the heavier isotope (^2H/deuterium or ^{13}C) than occurs naturally. The tracer isotope is either located at specific sites in the molecule, or more or less uniformly distributed at all sites. Molecules which contain different distributions of isotopes, but are otherwise identical, are known as isotopomers.

Stable isotope tracers are either prepared by chemical synthesis or isolated from biomass grown on a labelled substrate. Chemical synthesis is used to produce site-specific labelling, and the preparation of fatty acids specifically labelled with deuterium has recently been reviewed.[1] The costs of some commercially available labelled fatty acids and other tracers used to study lipid metabolism are given in Table 2.

In addition to commercially available tracers, there are a few published reports of biomass-derived ^{13}C-labelled fatty acids and triacylglycerols. Microalgae grown on [1-^{13}C]glucose produced triacylglycerols rich in 22:6($n-3$) which contained between 2 and 7 atom percent ^{13}C.[2,3] The fungus *Mortierella alpina* was grown on [^{13}C]glucose and produced 20:4($n-6$) with around 80 atom percent ^{13}C.[4] The labelling pattern of this product was not uniform, and depended on the labelling pattern of the glucose substrate.

In some circumstances 'natural tracers' can be used, exploiting the small differences in ^{13}C content of different plant oils. Plants convert atmospheric carbon dioxide to carbohydrates during photosynthesis. In higher plants, two different mechanisms of photosynthesis are known, the 'C_3' route where a

Table 2 *Typical costs of stable isotope tracers used to study lipid metabolism*

Tracer	cost (£/g)	cost (£/g ^{13}C)
Deuterium oxide (D$_2$O)	5	
Palmitic acid - d$_{31}$	170	
Sodium acetate (1-^{13}C)	56	357
Palmitic acid (1-^{13}C)	72	1400
Palmitic acid (1,2,3,4-^{13}C)	800	4000
Oleic acid (U-^{13}C)	1600	1900
Linoleic acid (U-^{13}C)	1600	1900

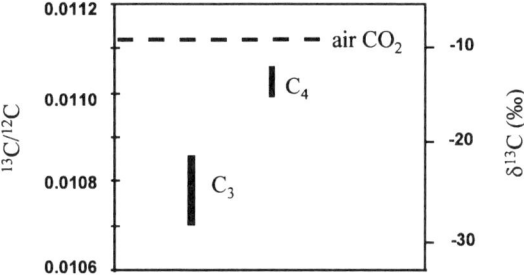

Figure 1 *Approximate natural range of $^{13}C/^{12}C$ ratio in C_3 and C_4 plants. The range of $\delta^{13}C$ (more conveniently used for natural abundance differences) is also shown*

three-carbon intermediate is the first product and the 'C$_4$' route where this is a four-carbon compound. Atmospheric carbon dioxide has a constant ^{13}C/^{12}C ratio, and the C$_3$ pathway fixes ^{13}CO$_2$ more slowly relative to ^{12}CO$_2$ than does the C$_4$ pathway. The result is that C$_3$ plants (most cereals, potatoes, sugar beet) have a lower ^{13}C/^{12}C ratio than C$_4$ plants such as maize and sugar cane (Figure 1). Further fractionation occurs during the biosynthesis of proteins and lipids, such that they (particularly lipids) have lower ^{13}C/^{12}C ratios than carbohydrates from the same source.

The only commercial seed oil derived from a C$_4$ plant is corn (maize) oil, with a higher ^{13}C content than all other dietary fats. Corn oil fatty acids can be regarded as very lightly enriched tracers if the diet is first controlled to avoid C$_4$ components. When C$_4$-derived lipid from corn oil is subsequently taken, the metabolic fate of the corn oil fatty acids can be traced with the sensitive analytical methods now available.[5]

Measuring Stable Isotopes

Stable isotope content can be measured either as an absolute amount or as the ratio of the minor isotope to the naturally abundant isotope. Isotope ratio measurements are the most common, and afford the possibility of precise quantification. The most common techniques for isotope ratio measurement use mass spectrometry, where molecular or fragment ions of different mass-to-charge ratio can be detected and quantified. For detecting tracer enrichments of 1% or more mass spectrometer systems such as GC-MS or liquid chromatography-MS (LC-MS) may be used. These systems combine separation with isotope ratio measurement. Site-specific information about the location of labelled atoms and/or an indication of the number of labelled atoms present can be obtained from each component.

NMR detects specific nuclei, ^1H, ^2H, ^{13}C and ^{17}O being of particular relevance in lipid compounds. Isotope ratios are not measured directly, but may be estimated indirectly on occasions. The strength of NMR lies in measuring the distribution of stable isotopes at different molecular positions of intact molecules without the need for degradative chemistry. ^{13}C enrichments of several percent are needed for NMR analysis, making this impractical for many human studies. However, data from model systems using ^{13}C NMR of multiply or uniformly labelled tracers can complement those obtained by mass spectrometry.[6]

A highly specialised field of mass spectrometry has developed for precise isotope ratio measurement, known as isotope ratio mass spectrometry (IRMS). This technique is essential for studies of natural variation and for detecting low levels of tracer. IRMS is restricted to low molecular weight stable gases, and other compounds must first be converted to a suitable gas. The recent development of GC-C-IRMS combines compound separation, conversion to a measurable gas and isotope ratio measurement.

Terms and Units

A number of terms and units are used to describe the amount of stable isotope in a sample. In tracer studies, stable isotope labelled molecules are added to the system and the mole fraction of added tracer in different compartments is used to calculate flux or synthetic rates. The mole fraction of stable isotope tracer is variously referred to as enrichment, mole or atom % excess (MPE or APE) and is equivalent to the 'specific activity' used in tracer studies with radio-isotopes. Stable isotope enrichment can be calculated from isotope ratio measurements alone. This contrasts with radio-isotope tracers where the specific activity is calculated from separate measurements of radioactivity and concentration. Only when the absolute amount of stable isotope tracer in a system is required are both concentration and isotope ratio measurements needed.[7]

All mass spectrometer based methods measure the enrichment as the difference between the natural abundance isotope ratio and the isotope ratio in the enriched sample. The mass spectrum of the molecule labelled with a stable isotope is essentially the same as that of the unlabelled molecule, but shifted up by one or more mass units. In the natural abundance spectrum, molecular or fragment masses contain isotopic contributions from all the elements present, which are seen at [M+1] and higher masses. The natural abundance contribution from a particular element increases as the number of atoms of that element in the ion increases. In CO_2, the M+1 ratio of the molecular ion is ~1.1%, close to the natural abundance value of ^{13}C as the contribution from

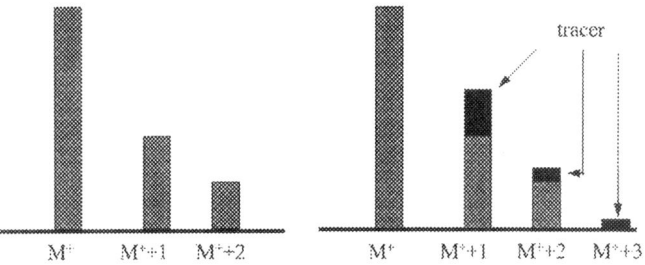

Figure 2 *Isotopomer intensities for an unlabelled molecular or fragment ion (left) and the same signal for a mixture of unlabelled compound and tracer labelled with a single heavy isotope which is one mass unit heavier than the light isotope (right)*

^{17}O is small. For a C_{18} methyl ester molecular ion, the M+1 ratio is ~21% again due mainly to the ~1.1% probability of ^{13}C being in each of the 19 carbons. Isotopic contributions from ^{2}H and ^{17}O are small. In mixtures of labelled and unlabelled compounds, the two mass spectra are superimposed (Figure 2).

A number of more or less rigorous methods are used to calculate enrichment from measured mass spectra. The abundance (A) of the enriched (enr) and unenriched or baseline (bas) samples can be calculated from their isotope ratios (R):

$$A = \frac{R}{R+1} \quad (1)$$

and the enrichment calculated by subtraction:

$$APE = 100 \times (A_{enr} - A_{bas}) \quad (2)$$

These relationships are only strictly true when there are no contributions from higher isotopomers, but in many cases (and for enrichments below ~10%) this does not produce significant errors.

The relationship

$$APE = \frac{100 \times (R_{enr} - R_{bas})}{R_{enr} - R_{bas} + 1} \quad (3)$$

based on the measured M+1/M ratios may be used with a similar caveat.[8]

A number of methods have been used to obtain linear relationships to improve the precision and accuracy of enrichment calculations.[9-11] The interpretation of overlapping mixtures of differently labelled compounds has been approached in a number of ways.[12-15] These methods use data from all the measurable isotopomer peaks. The correction for natural abundance may be made by calculating the expected values rather than by measuring unenriched samples.[16] Computer programs are available to carry out some of these algebraically complex calculations.[11,16] The choice of calculation method depends on the quality of the data, the particular application and the complexity of the model used to interpret the data.

Isotope ratios are not a convenient way to express the small differences in isotopic enrichment measured by IRMS. The 'delta notation' (δ) expresses the difference between the isotope ratio of a sample and that of a defined standard as a fraction of the standard's isotope ratio. The value is usually measured in parts per thousand or 'per mil' (‰):

$$\delta(‰) = \frac{1000 \times (\text{isotope ratio of sample} - \text{isotope ratio of standard})}{\text{isotope ratio of standard}} \quad (4)$$

The working standard used during the analysis is ultimately related to an International Standard curated by the International Atomic Energy Agency in Vienna. The International Standard for hydrogen is V-SMOW, with an $^2H/^1H$ ratio of 0.00015576 ($\delta^2H_{V\text{-}SMOW} = 0‰$), and that for carbon is V-PDB with a $^{13}C/^{12}C$ ratio of 0.0112372 ($\delta^{13}C_{V\text{-}PDB} = 0‰$).

When enrichments are calculated from IRMS data, the sample isotope ratio is calculated from the measured δ and the isotope ratio (R) of the International Standard before use in the above equations:

$$R_{sample} = R_{standard} \times \left(\frac{\delta_{sample}}{1000} + 1\right) \quad (5)$$

Corso and Brenna[17] have recently proposed the use of the relative isotope fraction (ϕ) rather than δ for IRMS tracer measurements. The relative isotope fraction is calculated in an analogous way to δ, but is expressed relative to the isotopic abundance (A) of the International Standard rather than its isotope ratio:

$$\phi = \left(\frac{A_{sample} - A_{standard}}{A_{standard}}\right) \times 1000 \quad (6)$$

Mass balance calculations using ϕ are accurate even when enrichments are well removed from the natural abundance standard, which is not true for those using δ.

3 Analytical Instrumentation

GC-MS

GC-MS systems link gas chromatographic separation with (usually) quadrupole mass spectrometry of each eluting component of the injected mixture (Figure 3). GC-MS methods are appropriate for ^{13}C and ^2H tracer measurements when there is ~1% or more tracer in the analyte, and offer compound-specific isotope analysis along with the opportunity to obtain structural confirmation on each separated component. Isotope ratio measurements are carried out on nanogram to picogram samples using selected-ion monitoring (SIM) to optimise the precision by collecting data only on the ions of interest.

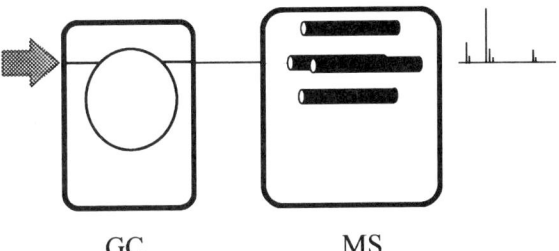

Figure 3 *Schematic layout of GC-MS. Eluent from the GC column is carried directly to the MS ion source*

Satisfactory measurements are possible using bench-top quadrupole GC-MS as well as higher performance instruments. The technique can be pushed to lower enrichments (~0.1%) if care is taken in sample isolation and purification, but cannot approach the isotopic sensitivity and precision of IRMS measurements. Systematic errors in measured isotope ratios occur if insufficient MS scans are collected per GC peak.[18-20] Using SIM and collecting 10 or more scans per peak, a bias of less than 1% can be achieved.[20]

Methyl esters are the usual derivatives for measuring fatty acid enrichment, although these can show variation in measured isotope ratio with sample size. This is due to hydrogen transfer reactions in the ion source resulting in an $[M+H]^+$ ion which contributes to the M+1 signal (Figure 4), and this reaction depends on the amount of sample in the ion source.[21] The problem can either be corrected for[22] or avoided by using a constant sample size and calibration standards. Further, no sample size effect was observed when methyl esters were analysed using a magnetic sector instrument, but one was found using a quadrupole analyser.[23] Pentaflurobenzoyl esters have been used as an alternative derivative, and did not show sample size dependence of the measured isotope ratio.[24]

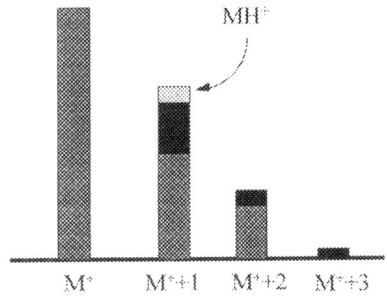

Figure 4 *Interfering contribution of MH^+ to M+1 signal of fatty acid methyl esters owing to ion–molecule reaction in the ion source*

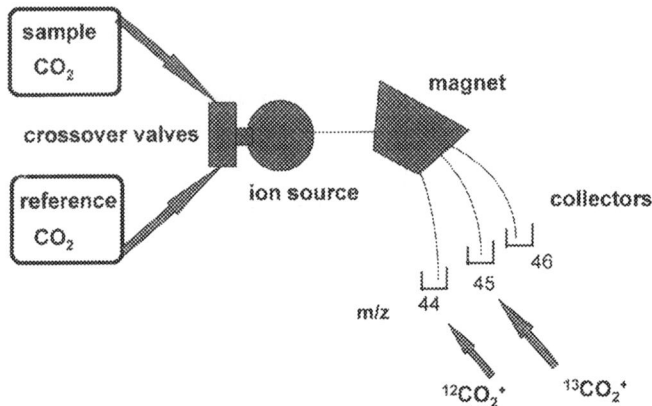

Figure 5 *Schematic diagram of a DI-IRMS. Sample and reference gases are fed to the cross-over valves by long capillaries to ensure equal signals and avoid fractionation*

Isotope Ratio Mass Spectrometry

Tracer levels of 1% or more can be measured using a wide range of mass spectrometers, but lower levels (down to 0.001%) and natural abundance variations can only be measured using dedicated isotope ratio mass spectrometers (IRMS). While IRMS are capable of impressive accuracy and precision in isotope ratio measurements, they are extremely restricted in the compounds that they can measure. Light elements such as hydrogen and carbon are measured as stable low-molecular weight gases (H_2 and CO_2 most commonly). This means that other analytes such as fatty acids must be converted to one of these before the isotope ratio measurement can be made. Until recently this was a major impediment in the application of stable isotope methods to lipids, but the development of systems combining sample conversion or both sample separation and conversion with the IRMS have greatly increased the accessibility of measurements on lipids and individual fatty acids.

Dual Inlet-IRMS. Dual inlet-IRMS (DI-IRMS) are capable of the highest precision and accuracy of isotope ratio measurements on pure gases such as CO_2 and H_2. In these instruments, samples of analyte gas and reference gas are contained in reservoirs connected by carefully matched capillaries to a magnetic sector mass spectrometer (Figure 5). The mass spectrometer has an electron-impact ion source and three or more spatially separate Faraday cup collectors to measure the isotopomer ion currents simultaneously. The pressures of the sample and reference gases are precisely matched to give equivalent signals. Sample and reference gas then enter the mass spectrometer alternately for periods of 10 s or more while their isotope ratios are measured. This process may be repeated for 10 or more cycles. Under optimum conditions, precisions of >0.01‰ are possible for $\delta^{13}C$ and >0.1‰ for δ^2H.

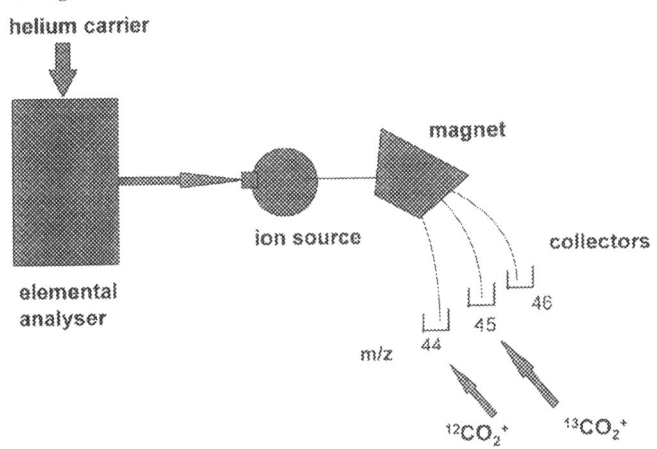

Figure 6 *Schematic diagram of a CF-IRMS. Samples are converted to CO_2 in the elemental analyser. The CO_2 is carried to the ion source by the helium carrier*

The limiting factor in applying DI-IRMS to lipid compounds is the need to isolate and convert milligram amounts of sample to hydrogen or carbon dioxide gas. Despite the inherent difficulties, this has been the only practical method for measuring low deuterium enrichments in triacylglycerols and cholesterol for studies of lipid synthesis. In a time consuming process, the triacylglycerol or cholesterol is isolated (free of any hydrogen-containing impurities or solvents), oxidised to water by sealed tube combustion, the water collected and reduced to hydrogen gas by hot-metal reduction before eventual IRMS measurement.[25,26]

Continuous Flow-IRMS. Continuous flow-IRMS (CF-IRMS) links the conversion of the sample to a suitable gas with the IRMS measurement, using a helium carrier gas to sweep the analyte gas into the IRMS (Figure 6). The most common sample converter is a Dumas combustion elemental analyser, which converts all sample carbon to CO_2. The same mass analyser is used as for DI-IRMS, but measurements are made on transient peaks from the elemental analyser which are a minute or so wide. Standards are introduced before and after sets of up to 10 samples, and despite the reduced frequency of sample/reference comparisons, the precision is not greatly reduced. A precision of 0.1‰ is possible for $\delta^{13}C$ measurement on samples containing 100–200 µg C. These systems are fully automated and can process batches of more than 100 samples per day. However, for studies of lipid molecules there is still the problem of isolating sufficient pure material free of extraneous carbon-containing solvents.

Continuous flow systems are routinely used for measuring ^{13}C in expired breath, using automated sampling and GC to extract and purify CO_2 from disposable sample tubes. These rapid analyses are used for breath tests of metabolic function, or to measure rates of substrate (*e.g.* fatty acid) oxidation

Until recently, continuous-flow systems for hydrogen were not available,

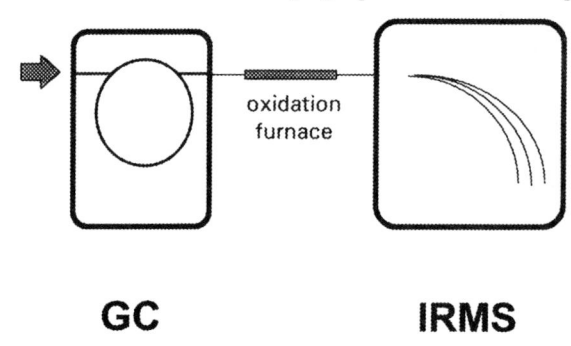

Figure 7 *Schematic diagram of a GC-C-IRMS. Components eluting from the GC are converted to CO_2 in the micro-oxidation furnace. The CO_2 is carried to the ion source by the helium carrier*

and all hydrogen IRMS used dual-inlet instruments. Recent developments in continuous-flow hydrogen analysis are described below.

Gas Chromatography-Combustion-IRMS. Gas chromatography-combustion-IRMS (GC-C-IRMS) was developed from the CF-IRMS systems described above. The elemental analyser is replaced by a gas chromatograph to separate individual compounds, a micro-oxidation furnace to convert these to pulses of CO_2 and a water trap to remove water produced during oxidation (Figure 7). Peak broadening during the combustion is minimised by avoiding dead volumes and using makeup-flow to compensate for the inevitable increased diameter of the oxidation tube. The result is a series of CO_2 peaks entering the IRMS which closely parallel a flame ionisation detector trace of the mixture. GC-C-IRMS is a major step forward in compound-specific analysis in two respects. It allows analysis of a wide range of compounds, and requires only nanogram amounts of sample.

This new technique has already acquired an unfortunate variety of names and acronyms: compound-specific isotope analysis, stable isotope ratio monitoring-GC/MS, GC-combustion-IRMS or just GC-IRMS. GC-C-IRMS is used here as it describes succinctly the whole system. Although first described in 1984,[27] GC-C-IRMS systems have only been commercially available since the early 1990s. The three major IRMS manufacturers now offer GC-C-IRMS as standard options for their instruments. Brand has recently reviewed isotope ratio monitoring IRMS including GC-C-IRMS instrumentation and applications.[28]

In common with most GC analysis of fatty acids, GC-C-IRMS is usually carried out using methyl ester derivatives. Most samples will not have originated in this form, and the methyl ester carbon will come from a different source from the sample. The combustion step converts all the carbon in the analyte peak into CO_2 resulting in contamination or dilution of the carbon originally in the fatty acid. If the isotopic composition of the derivative carbon is known, then this can be corrected by a mass balance calculation. Isotope fractionation is not expected in the preparation of methyl esters of fatty acids. Rieley has discussed derivatisation of compounds for GC-C-IRMS at length and includes an analysis of the errors likely to be propagated in carrying out the mass balance calculation.[29]

Current Developments in GC-IRMS

GC-C-IRMS is now well established for carbon isotope measurement. At present, similar analysis of hydrogen is not possible. However, compound-specific IRMS for deuterium labelling would have immediate application in measuring fatty acid and cholesterol synthesis by labelled water. At present, the deuterium analysis is done by lengthy off-line methods. The initial stumbling block to developing a GC-IRMS method was the difficulty in measuring hydrogen gas in a continuous flow system. Excess helium carrier (m/z 4) interferes with the measurement of the very small signal for HD (m/z 3), due to imperfect focusing of the ion beams in a conventional IRMS analyser. This problem has now been resolved by two different approaches, a palladium filter system which transmits hydrogen but not helium[30] and a modified IRMS with high dispersion to widely separate the m/z 3 and m/z 4 ion beams.[31] Using a reduction or pyrolysis furnace, satisfactory results have now been obtained for continuous flow measurements on water[32] and a number of organic compounds.[33,34] As with DI-IRMS, correction is required for H_3^+ formation in the ion source. When compounds elute from a GC column there are further confounding effects due to separation of deuterium-containing isotopomers. This causes additional difficulties in accurately quantifying the isotope ratio, which have not yet been fully resolved. The ability to measure modest deuterium enrichments by GC-IRMS should be realised shortly, with natural abundance precision following the development of adequate correction algorithms for the time-dependent factors.

GC-C-IRMS destroys any site-specific information within the compound as all carbons are converted to CO_2. A recent report by Corso and Brenna[17] describes a post-GC pyrolysis step, converting saturated methyl esters to a series of terminal alkenes and alkene esters. These are separated by a second GC and the effluent split between an ion trap MS and combustion IRMS. The former identifies each component and the latter measures the ^{13}C enrichment. By combining all this information it is possible to determine the enrichment at each position along the fatty acid chain. This offers a one-step instrumental solution to problems that have previously required extensive degradative chemistry.[35,36] So far, only saturated acids have been used, but the potential for detailed study of fatty acid elongation, retro-conversion and carbon recycling *in vivo* has been clearly demonstrated.

4 Applications

Stable isotopes are used to study lipid metabolism in a number of ways. The choice of approach depends on the problem to be solved and the kind of data which can be obtained. Interpreting the data requires a model of the metabolic system, and appropriate stable isotope data must be collected to provide sufficient information for the model to be meaningful. The model may constrain the study approach, for example requiring the tracer to be infused for a prolonged period while a particular metabolic state (*e.g.* fasted or fed) is maintained. Constant infusion studies produce precise quantitative data under the study conditions, but are clearly far removed from a normal free living state. On the other hand, feeding a small tracer dose along with normal food provides a more realistic view of the fate of dietary lipids, but requires much more complex compartmental modelling to interpret the results. Indeed, it may be difficult to construct a fully satisfactory model with the presently available knowledge of lipid metabolism.

Constant Infusion Methods

Constant infusion methods have been most widely used to date, to study how changing the metabolic state alters the net rate at which fatty acids enter the circulation. These use either a simple one-pool model or relatively simple variants thereof. Infused tracer is diluted as it enters the bloodstream, and the eventual steady-state dilution reflects how much unlabelled fatty acid is entering the same pool. The latter may come from digestion, or from release of fatty acid from storage or from *de novo* synthesis. This simple model cannot directly distinguish between these sources, but can monitor net changes with good precision.

Enrichment data has normally been collected by GC-MS, requiring plasma enrichments of several percent for accurate measurement of any changes. The labelled free-fatty acid must be infused bound to albumin to mimic the circulating unlabelled acid, and avoid toxic concentrations of unbound fatty acid. Human albumin is needed to avoid immunological problems and must be purified to remove any risk of infection with HIV or other blood-product-borne pathogens. The combined cost of albumin and tracer makes these expensive studies. GC-C-IRMS may be useful in future in greatly reducing the amount of fatty acid required, but the more expensive [U-^{13}C]-fatty acids may be preferred for GC-C-IRMS rather than [1-^{13}C]-labelled compounds used when analysing by GC-MS.

The steady state conditions are achieved more rapidly by giving a priming dose of tracer (equivalent to ~ 1 h infusion). Steady state can also be achieved in the bicarbonate pool (again following a priming dose) and the enrichment of expired breath carbon dioxide can be used to calculate the oxidation rate of the traced fatty acid.[37]

In practice, obtaining reliable data from constant infusion studies requires careful consideration of sampling site, and data interpretation may require models taking several body pools into account.[38,39] Goodenough and Wolfe have examined this with particular regard to the infusion of ^{13}C-labelled palmitate.[40]

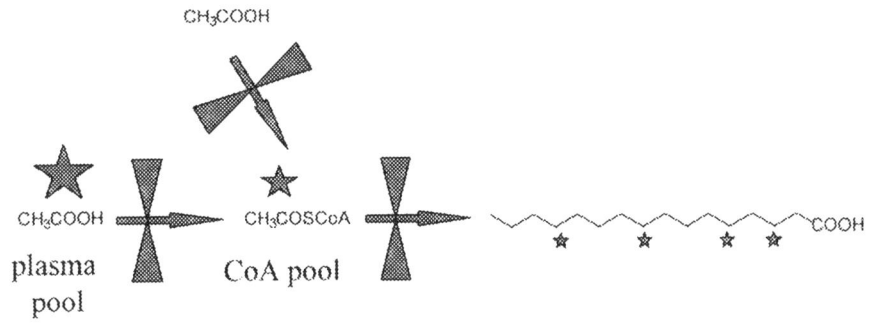

Figure 8 *Incorporation of label from acetate during the biosynthesis of fatty acids. The synthesis rate can be calculated from the enrichment of the acetyl CoA precursor pool and the eventual enrichment of the fatty acid product. The enrichment of acetate in the precursor pool is not the same as in plasma, owing to dilution by endogenous acetate supply*

Fatty Acid and Cholesterol Synthesis

Fatty acids and cholesterol are both biosynthesised by the condensation of acetate units. In principle, the rate of synthesis can be measured by following the incorporation of label from a precursor pool of known isotopic enrichment into the product. In practice, however, this is not straightforward. The precursor pool which must be measured is acetyl CoA. This is not a discrete accessible pool (Figure 8). The pools of acetyl CoA units for the synthesis of fatty acids, cholesterol and ketones, and for oxidation, differ in the liver and may respond differently to altered physiological states. Since the measurement of precursor enrichment is not open to direct attack, the problem must be approached indirectly. Two methods have been developed—mass isotopomer distribution analysis (MIDA) and labelled water incorporation. These methods have been reviewed thoroughly and informatively by Hellerstein.[41,42]

By regarding lipid synthesis as the polymerisation of acetate units, MIDA circumvents the problem of measuring the precursor enrichment. Instead, the precursor enrichment is calculated from the distribution of labelled units in the polymeric product. The distribution of label in the precursor (precursor enrichment) and in the product (MIDA) are linked statistically by a binomial or multinomial expansion. This relationship is used in reverse to calculate the precursor enrichment from the isotopomer distribution in the product, measured by GC-MS. In principle, the whole range of possible isotopomers can be measured, but in practice sufficient enrichment is seldom achieved beyond the mass corresponding to incorporation of two or three precursor units.

The labelled water method uses body water labelled with deuterium as a surrogate for the precursor enrichment. Water diffuses readily across cell membranes and is not compartmentalised. Water is the hydrogen source for reduction of carbonyl groups by NADPH during biosynthesis (Figure 9), and hydrogen incorporated into lipid molecules is not then exchanged with the

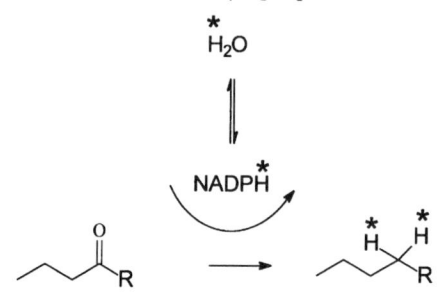

Figure 9 *Incorporation of labelled hydrogen from body water during the biosynthesis of fatty acids*

medium or other lipid molecules. The deuterium enrichment of body water is readily measured. Deuterium enrichment in the lipid products is measured by conversion to water as described above.[25,26] The need for IRMS methods arises from the toxicity of deuterium oxide at levels much above 1% in body water (causing initially vertigo and nausea). This limits the doses that can be given and the eventual enrichment of the product. Jones has recently reviewed the use of deuterated water incorporation for studying lipogenesis.[43]

Diraison *et al.* have recently reported the use of GC-MS to measure deuterium incorporation into fatty acids from labelled water studies.[44] They used deuterium oxide enrichments of ~0.3% of body water, three to six times greater than needed in studies using IRMS analysis. These doses were well tolerated and are considered safe in non-pregnant adults. GC-MS analysis allows measurement of an individual fatty acid or cholesterol without preparative isolation of these compounds. However, this approach is limited to subjects for which these higher doses are considered safe, and a GC-IRMS approach allowing low deuterium oxide doses and compound-specific analysis remains highly desirable.

MIDA and labelled water methods are complementary approaches to a common problem, and both are model dependent. Calculations of synthesis rates by MIDA or labelled water incorporation use computational models developed using a number of biological model systems. There remain some problems in being sure that parameters obtained from model systems are generally applicable and do not change with altered metabolic state. The complementary nature of these two methods allows some degree of cross-checking, and the construction of more robust models.

In particular, the number of deuterium atoms per carbon atom in the product must be known to calculate synthesis from deuterium incorporation. This has been studied recently using a rat model and a sufficiently high deuterium dose (~10%) to allow isotopomer distributions to be measured.[45] The isotopomer distributions in fatty acids and cholesterol resulting from incorporation of deuterium and ^{13}C from labelled glucose were investigated in cell culture.[46] The authors of this study emphasise the detail of information available on the economy of acetyl units and reducing equivalents that results from the use of stable isotopes to study the system.

GC-C-IRMS Studies

The increasing availability of ^{13}C-labelled tracers and sensitive compound-specific GC-C-IRMS detection has led to a rapid expansion of stable isotope studies of lipid metabolism *in vivo*. A small dose of labelled fatty acid can be followed through different metabolic pools, or through desaturation, elongation or degradation steps. The access to information on several different fatty acids in a single sample greatly enhances the power of this approach. Single dose studies can give a wealth of qualitative information. Converting this to quantitative results will require more complex compartmental models than currently used for constant infusion studies.

The first studies using singly- or [U-^{13}C]-labelled fatty acids and GC-C-IRMS concentrated on establishing the dose required to give useful information. Rats of 300 g weight fed 3 mg of triacylglycerol containing 36% uniformly labelled 22:6(n−3) containing 2 atom % ^{13}C gave a readily measurable enrichment in very-low-density lipoprotein triacylglycerols, low-density lipoprotein neutral lipids, high-density lipoprotein triacylglycerols and phospholipids.[2] Retro-conversion to 22:5(n−3) and 20:5(n−3) was also detected in the high-density lipoprotein phosphatidylcholine. The dose used corresponds to 0.04 mg/kg of pure 22:6, fully labelled, *i.e.* only ~2.5 mg for a human adult, and the analyses could be carried out on 10 ml plasma.

Enrichment detectable by GC-C-IRMS was found in the serum triacylglycerols of healthy human adults following a 100 mg oral dose of 1- or 8-[^{13}C]-triolein.[47] Enrichment was seen in palmitic, stearic, oleic and linoleic acids in the triacylglycerols but was not detected in the phospholipid or cholesterol ester fractions. A 10 mg dose of 42% uniformly labelled 18:0 in an adult human produced enrichments of ~0.1 APE in the chylomicron fraction and ~0.01 APE in both 18:0 and 18:1 in the dense fraction of plasma after 4 h.[48]

The metabolic ability of neonates and infants to synthesise and metabolise dietary fatty acids is an area likely to benefit from GC-C-IRMS methods. Already a number of studies have looked at the production of long-chain PUFA from dietary linoleate and linolenate. Pre-term infants given ^{13}C labelled 18:2 and α-18:3 in formula feed for 48 h produced highly labelled 20:4 and 22:6 in plasma phospholipids.[49] These data do not support the hypothesis that reduced Δ^6-desaturase activity is the main factor leading to low 20:4 and 22:6 levels in premature infants. As mentioned above, GC-C-IRMS is sufficiently sensitive for lipids with different natural abundances of ^{13}C to be used as tracers. Replacing C_3-derived lipid with corn oil (C_4) in infant feed produced a progressive enrichment in the serum linoleic acid.[5] Progressive enrichment was seen also in dihomo-γ-linolenic and arachidonic acids, demonstrating active chain elongation and desaturation. Using a simple model, it was estimated that 23% of plasma arachidonic acid originated from dietary linoleic acid by day 4 of the study.

We can look forward to having information on fatty acid metabolism which could not have been obtained previously. The compartmental metabolism[3] and retro-conversion[50] of 22:6(n−3) have been documented in rats and humans following a single dose of ^{13}C-labelled 22:6($n-3$). 22:6($n-3$) appeared rapidly (over 2–3 h) in triacylglycerols in the very-low-density lipoprotein and chylomicron fractions. Up to 12 h post-ingestion, most labelled 22:6($n-3$) bound to albumin was unesterified, but thereafter, the lysophosphatidylcholine fraction carried higher ^{13}C concentrations. Retro-conversion to 22:5($n-3$) and 20:5($n-3$) was 9% of total plasma 22:6($n-3$) in rats but only 1.4% in humans.

This field is now growing rapidly. GC-C-IRMS can probe areas of lipid metabolism not previously amenable to direct or detailed study. Many of the reports to date have been exploring these new areas, showing what can be done. The examples above demonstrate the sensitivity of GC-C-IRMS in following the metabolic fate of fatty acids in animal models and in relatively non-invasive studies in humans. Detailed interpretation of the results is still at an early stage, and the there is a need to develop experimental protocols that ensure adequate data are available to produce robust results.

Acknowledgement

This review is published as part of a programme funded by the Scottish Office Agriculture, Environment and Fisheries Department.

References

1. L. Crombie, *Synthesis in Lipid Chemistry*, ed. J. H. P. Tyman, Royal Society of Chemistry, Cambridge, 1996, p. 34.
2. N. Brossard, C. Pachiaudi, M. Croset, S. Normand, J. Lecerf, V. Chirouze, J. P. Riou, J. L. Tayot and M. Lagarde, *Anal. Biochem.*, 1994, **220**, 192.
3. M. Croset, N. Brossard, C. Pachiaudi, S. Normand, J. Lecerf, V. Chirouze, J. P. Riou, J. L. Tayot and M. Lagarde, *Lipids*, 1996, **31**, S109.
4. H. Kawashima, K. Akimoto, T. Fujita, H. Naoki, K. Konishi and S. Shimizu, *Anal. Biochem.*, 1995, **229**, 317.
5. H. Demmelmair, U. v-Schenck, E. Behrendt, T. Sauerwald and B. Koletzko, *J. Pediatr. Gastroenterol. Nutr.*, 1995, **21**, 31.
6. S. C. Cunnane and S. S. Likhodii, *Can. J. Physiol. Pharmacol.*, 1996, **74**, 761.
7. I. M. Campbell, *Bioorg. Chem.*, 1974, **3**, 386.
8. W. T. Buckley, S. N. Huckin and G. K. Eigendorf, *Biomed. Mass Spectrom.*, 1985, **12**, 1.
9. B. N. Colby and M. W. McCaman, *Biomed. Mass Spectrom.*, 1979, **6**, 225.
10. D. A. Schoeller, *Biomed. Mass Spectrom.*, 1976, **3**, 265.
11. L. M. Thienpont, B. Van Nieuwenhove, D. Stöckl and A. P. De Leenheer, *J. Mass Spectrom.*, 1996, **31**, 1119.
12. E. D. Bush and W. F. Trager, *Biomed. Mass Spectrom.*, 1981, **8**, 211.
13. K. Korzekwa, W. N. Howald and W. F. Trager, *Biomed. Environ. Mass Spectrom.*, 1990, **19**, 211.
14. J. A. Vogt, T. E. Chapman, D. A. Wagner, V. R. Young and J. F. Burke, *Biol. Mass Spectrom.*, 1993, **22**, 600.
15. W. Schramm, W. Schill and T. Louton, *Biomed. Mass Spectrom.*, 1979, **6**, 335.
16. C. A. Fernandez, C. Des Rosiers, S. F. Previs, F. David and H. Brunengraber, *J. Mass Spectrom.*, 1996, **31**, 255.

17. T. N. Corso and J. T. Brenna, *Proc. Natl. Acad. Sci. USA*, 1997, **94**, 1049.
18. B. R. Pettit, *Biomed. Environ. Mass Spectrom.*, 1986, **13**, 473.
19. D. E. Matthews, K. B. Denson and J. M. Hayes, *Anal. Chem.*, 1978, **50**, 681.
20. D. E. Matthews and J. M. Hayes, *Anal. Chem.*, 1976, **48**, 1375.
21. B. W. Patterson and R. R. Wolfe, *Biol. Mass Spectrom.*, 1993, **22**, 481.
22. E. A. Bergner and W.-N. P. Lee, *J. Mass Spectrom.*, 1995, **30**, 778.
23. M. G. Kienle and F. Magni, *Biol. Mass Spectrom.*, 1994, **23**, 173.
24. J. Zamecnik, C. Fayolle and A. Vallerand, *Biol. Mass. Spectrom.*, 1994, **23**, 804.
25. C. A. Leitch and P. J. H. Jones, *Biol. Mass Spectrom.*, 1991, **20**, 392.
26. C. A. Leitch and P. J. H. Jones, *J. Lipid Res.*, 1993, **34**, 157.
27. A. Barrie, J. Bricout and J. Koziet, *Biomed. Mass Spectrom.*, 1984, **11**, 583.
28. W. A. Brand, *J. Mass Spectrom.*, 1996, **31**, 225.
29. G. Rieley, *Analyst*, 1994, **119**, 915.
30. H. J. Tobias, K. J. Goodman, C. E. Blacken and J. T. Brenna, *Anal. Chem.*, 1995, **67**, 2486.
31. S. J. Prosser and C. M. Scrimgeour, *Anal. Chem.*, 1995, **34**, 1992.
32. I. S. Begley and C. M. Scrimgeour, *Rapid Commun. Mass Spectrom.*, 1996, **10**, 969.
33. H. J. Tobias and J. T. Brenna, *Anal. Chem.*, 1996, **68**, 3002.
34. I. S. Begley and C. M. Scrimgeour, *Anal. Chem.*, 1997, **69**, 1530.
35. D. K. Monson and J. M. Hayes, *J. Biol. Chem.*, 1980, **255**, 11435.
36. D. K. Monson and J. M. Hayes, *Geochim. Cosmochim. Acta*, 1982 **46**, 139.
37. L. S. Sidossis, A. R. Coggan, A. Gastaldelli and R. R. Wolfe, *Am. J. Physiol.*, 1995, **269**, E649.
38. D. K. Layman and R. R. Wolfe, *Am. J. Physiol.*, 1987, **253**, E173.
39. S. L. Lehman and W. C. Stanley, *Am. J. Physiol.*, 1988, **255**, E94.
40. R. D. Goodenough and R. R. Wolfe, *Am. J. Clin. Nutr.* 1983, **37**, 1004.
41. M. K. Hellerstein, *Curr. Opin. Lipidol.*, 1995, **6**, 172.
42. M. K. Hellerstein, *Lipids*, 1996, **31**, S117.
43. P. J. H. Jones, *Can. J. Physiol. Pharmacol.*, 1996, **74**, 755.
44. F. Diraison, C. Pachiaudi and M. Beylot, *J. Mass Spectrom.*, 1997, **32**, 81.
45. F. Diraison, C. Pachiaudi and M. Beylot, *Metab. Clin. Exp.*, 1996, **45**, 817.
46. W. N. P. Lee, L. O. Byerley, S. Bassilian, H. O. Ajie, I. Clark, J. Edmond and E. A. Bergner, *Anal. Biochem.*, 1995, **226**, 100.
47. C. C. Metges, K. Kempe and G. Wolfram, *Biol. Mass Spectrom.*, 1994, **23**, 295.
48. K. J. Goodman and J. T. Brenna, *Anal. Chem.*, 1992, **64**, 1088.
49. V. P. Carnielli, D. J. L. Wattimena, I. H. T. Luijendijk, A. Boerlage, H. J. Degenhart and P. J. J. Sauer, *Pediatr. Res.*, 1996, **40**, 169.
50. N. Brossard, M. Croset, C. Pachiaudi, J. P. Riou, J. L. Tayot, and M. Lagarde, *Am. J. Clin. Nutr.*, 1996, **64**, 577.

3
Trans Unsaturated Fat in Health and Disease

David Kritchevsky

THE WISTAR INSTITUTE, 3601 SPRUCE STREET, PHILADELPHIA, PA 19104, USA

1 Introduction

In most naturally occurring unsaturated fats the double bonds are in the *cis* configuration. However, *trans* double bonds do occur naturally in a number of plants,[1] in the body fat of ruminants, and in milk. The major *trans* fatty acid of milk and beef fat is vaccenic acid (11*t*-18:1). The principal source of *trans* fat in the diet of the developed world is as a component of partially hydrogenated fat. Partial hydrogenation of vegetable or marine oils yields fats with greater stability, which can be used directly as margarines, salad oils, cooking oils and shortenings or used in the preparation of other foods. Heat treatment or frying of vegetable oils also yields some fatty acids containing *trans* unsaturated double bonds but the yields are quite small.[2-4] Deodorization of oils may also yield a very small amount of *trans* fat.[5]

Commercial hydrogenation of vegetable oils results in mostly monoenoic fats and most of the component *trans*-unsaturated (*trans*) fatty acids are present in the monoenoic fraction. The proportion of *trans* fats can be reduced by altering conditions of hydrogenation. In the course of the hydrogenation process the double bonds may also migrate along the hydrocarbon chain, yielding monoene fatty acids with double bonds anywhere from carbon 4 to carbon 16 (Table 1).[6] Few biological studies which have been carried out using *trans* fats have used specific positional isomers (usually elaidic acid); insofar as the rest are concerned we must still clear up the possibility that the results may be due to specific positional isomers.

The presence of a *cis* double bond in a fatty acid confers a bend to the acyl chain resulting in a flexible molecule. Fatty acids containing *trans* double bonds are linear because the angle conferred by the double bond is much smaller than that seen in *cis* fatty acids. Consequently, *trans* fatty acids are more rigid in structure and higher melting than their *cis* counterparts and

Table 1 Double bond position of monoenes in hydrogenated fat[a]

Fat, %18:1	Double bond position (%)					
	4–7	8	9	10	11	12–16
44.8 cis	0.4	1.3	39.2	1.5	1.2	1.2
21.9 trans	2.0	2.8	3.9	4.7	3.9	4.6
19.8 cis	0.4	1.0	12.7	2.0	1.7	2.1
23.8 trans	1.2	2.8	5.0	5.4	4.5	5.0
19.3 cis	1.3	2.1	6.9	3.2	2.6	3.3
45.8 trans	4.1	5.9	9.8	9.3	7.7	9.3

[a] After Dutton.[6]

resemble saturated fatty acids. Since spatial configurations of *cis* and *trans* fatty acids differ, one might expect differences in the chemical and biological properties. The melting point of oleic acid (9c-18:1) is 13 °C, that of elaidic acid (9t-18:1) is 44 °C and that of stearic acid (18:0) is 70 °C. The biological properties of the *trans* fats were of interest and concern from the very first. In 1954, Melnick and Deuel[7] reported the 'iso-oleic' acids, as the *trans* fats were then called, could be used as nutrients and concluded that 'hydrogenated fats compare favorably with a natural fat of comparable firmness'. When fed as part of an adequate fat regimen (*i.e.*, sufficient essential fatty acids) *trans* fats do not affect growth,[8] but when fed as the sole source of fat they exaggerate symptoms of essential fatty acid (EFA) deficiency.[9] A multigenerational study which involved feeding a margarine containing 35% *trans* fatty acids for a number of years showed no deleterious effects of growth or reproductive capacity of rats.[10] Alfin-Slater and Aftergood[11] fed rats diets containing 15% of six different fats with varying levels of *trans* fat for four generations and found no untoward effects on growth (Table 2). The rats from different groups exhibited similar patterns of reproductive capacity and mortality. The findings have been confirmed.[12] The overall conclusion from these data is that in the presence of sufficient levels of EFA *trans* fats do not affect growth or reproductive capacity of rats. Zevenbergen *et al*.[13] found that *trans* fats did not have an unfavorable effect on mitochondrial function provided there was sufficient linoleic acid present in the diet. Saturated fat fed in the absence of EFA would also be expected to have deleterious effects on metabolism and, indeed, it has been suggested[14,15] the *trans* fats be regarded as quasi saturated fats. Vergroesen[16] found that human volunteers fed elaidic acid exhibited serum cholesterol levels intermediate between those of subjects fed oleic acid or saturated fat.

Metabolism

In general, *cis* and *trans* fats are metabolized in a similar fashion. It should be pointed out that most of the data on *trans* fat metabolism have been obtained using test fats containing one double bond in the *trans* configuration. Fatty

Table 2 Influence of dietary trans fat on growth of rats[a]

| | | % Trans | | Avg. wt. (g)[b] | | | |
| | | | | Generation I | | Generation IV | |
Fat	18:1t	18:2t,t	18:2c,t	M	F	M	F
I	0.5	0	0	316	209	308	191
II	23.2	2.1	1.1	295	186	294	193
III	50.2	2.7	0	331	205	300	209
IV	27.3	3.0	3.8	319	193	298	181
V	12.2	2.3	0	340	200	323	183
VI	17.9	3.2	3.5	335	205	292	187

[a] After Alfin-Slater and Aftergood.[11]
[b] Twenty females, 12 males in each group.

acids containing two *trans* double bonds can have different and possibly injurious effects but they are not a common dietary component.

Trans fats are transported much like their *cis* counterparts. Pigs fed *cis* or *trans* isomers carry the particular fatty acid in all plasma lipoprotein fractions.[17] Rats fed *cis* or *trans* fat exhibit similar levels of plasma apolipoproteins A1 and B when the diet is free of cholesterol, but when fed *cis* or *trans* fat plus cholesterol the plasma of rats fed *trans* fat contains more apolipoprotein A1.[18] It has been shown repeatedly that dietary *trans* fatty acids are deposited in most body tissues (references 19–22 are representative). Generally, the incorporation of *trans* fats into tissues is proportional to the amount present in the diet[23] and when *trans* fat stimulus is removed the *trans* fatty acids disappear from tissue.[23,24] This is further evidence that *trans* fats are metabolized normally. Wood[22] fed rats hydrogenated soybean oil (33.4% *trans*, 66–6% *cis* isomers) and found the distribution of *cis* and *trans* octadecenoates shown in Table 3.

Table 3 Distribution of isomeric octadecenoates in tissues of rats fed partially hydrogenated safflower oil (33.4% trans)[a]

| | Triglycerides | | Phosphatidyl choline | |
	% Cis	% Trans	% Cis	% Trans
Brain	–	–	98	2
Heart	73	27	50	50
Kidney	76	24	67	34
Liver	85	15	45	55
Lung	69	31	73	27
Muscle	72	28	73	28
Spleen	71	29	58	42
Adipose tissue	66	34	58	42

[a] After Wood.[22]

Carboxyl labeled (^{14}C) oleic and elaidic acids are oxidized to $^{14}CO_2$ at the same rate.[25] In rats, carboxyl-labeled palmitic, oleic, and elaidic acids were oxidized and excreted into the lymph at the same rate. Carboxyl-labeled stearic acid was metabolized a little more slowly.[26,27]

Isolated rat heart mitochondria utilize the CoA esters of *trans*-monoenoic acids as substrates for β oxidation but the rate of oxidation is slower than that for the corresponding *cis* isomers.[28,29]

Fatty acid desaturating and elongating activity is slower when the substrate is *trans* fat.[30–32] Liver lecithin-cholesterol acyl *trans*ferase (LCAT) activity is unchanged in monkeys fed 3–6% *trans* fat[24] but is reduced in rats fed high levels of *trans* fat.[33,34] Cholesterol esters of *trans* fatty acids are synthesized[35] and hydrolysed[36] more slowly than those of *cis* fatty acids. Activity of hepatic microsomal, mitochondrial, or cytosolic enzymes in rabbits fed 3–6% *trans* fats were the same as those observed in rats fed a control diet.[37] *Trans* fats influence the production of eicosanoids but the results can be modulated by the type of dietary protein. Koga *et al.*[38] fed rats diets containing *cis* or *trans* fat and casein or soy protein. Prostaglandin E2 levels (pg/mL plasma) were 7.26 and 15.3 in rats fed *cis* fat and casein or soy protein, inclusion of *trans* fat in the diet has virtually no effect on the results. Leukotriene C4 production (ng/g spleen) was 43.5±6 and 36.0±3 in rats fed *cis* fat and casein or soy protein. Substitution of *trans* fat had no effect on leukotriene production in soy protein-fed rats but lowered that in the casein-fed groups to 28.1±3 ($p<0.05$).

Carcinogenesis

There have been few studies of the effects of *trans* fats on experimental carcinogenesis. Brown[39] fed diets containing 5 or 17% of saturated fats, *cis*-monoenes fats, polyunsaturated fats, or *trans* monoenes to mice that were also injected with 1,2-dimethylhydrazine (DMH). There were no differences among the dietary groups in liver tumor incidence. Selenskas *et al.*[40] compared the effects of corn oil, 'cis' fat (a blend containing 54.7% 18:1*c*) or 'trans' fat (a blend containing 38.3% 18:1*t* and 19.2% 18:1*c*) on mammary tumor incidence in rats fed 7,12-dimethylbenz[*a*]anthracene (DMBA). The fats were fed at levels of 5 or 20% and as Table 4 shows the *trans* fat had the least tumor promoting effect. In another comparison of native and hydrogenated (71% *trans* monoene) soybean oil on DMBA-induced mammary tumors in rats, the latter produced a slight reduction in tumor incidence and multiplicity.[41] Erickson *et al.*[42] fed the same fats as did Selenskas at the same levels to BALB/c mice bearing transplanted mammary tumors. They found no differences in latency or rate of primary tumor growth. When the cells were administered intravenously the *trans* fat was less effective than the *cis* fat in promoting blood-borne implantation and survival of the tumor cells. Reddy *et al.*[43] studied the effects of three levels of fat (23.5%, 13.6%, or 5%) on azoxymethane-induced colon tumors in rats. The fats were corn oil or fat mixtures providing 5.88% (low), 11.76% (intermediate) or 17.64% (high) *trans* fat. Tumor incidence in rats fed high, intermediate, or low levels of corn oil were 98, 63, and 67%, respectively. Incidence of colon tumors in rats fed high, intermediate or low levels of *trans* fat were 63, 67, and 57%, respectively. Watanabe *et al.*[44] compared the effects of olive oil (74.1% 18:1*c*) and partially hydrogenated corn oil (27.2% 18:1*c* and 42.0% 18:1*t*) on DMH-induced colon tumors in rats. Tumor incidence was virtually the same in the two groups, being 31.3% in the olive oil group and 35.3% in the rats fed the partially hydrogenated fat. Partially hydrogenated corn oil (39% 18:1*t*, 23% 18:1*c*) and a high oleic acid strain of safflower oil (62% 18:1*c*) were fed at a level of 5% to male and female rats of the Wistar-Furth-Osaka (colon cancer prone) strain. There were no differences in incidence of colon cancer (70.6% in *cis* group and 75.0% in *trans* group).[45] Linoleic acid is a growth factor for tumors[46] and saturated fats are generally less co-carcinogenic than unsaturated fats.[47] *Trans* monoene fats resemble saturated fats in relation to co-carcinogenicity.

Table 4 *Influence of cis or trans fat on experimental mammary carcinogenesis*[a]

Fat	% in diet	% Trans	Tumor incidence (%)
'Trans'[b]	5	1.5	16
	20	7.2	32
'Cis'[c]	5	–	24
	20	–	40
Corn oil	5	–	44
	20	–	80

[a] After Selenskas *et al.*[40]
[b] Blend containing 38.3% 18:1*t*, 19.2% 18:1*c*.
[c] Blend containing 54.7% 18:1*c*.

Experimental Atherosclerosis

The effects of *trans* fats on experimental atherosclerosis were of concern in the 1960s. McMillan's group at McGill University in Montreal, Canada, carried out three sets of studies in rabbits. The rabbits were fed 1 g of cholesterol and 6 g of fat for 12 weeks, at which time they were subjected to necropsy and their aortas examined for extent of atherosclerotic involvement. Table 5 bears a summary of the data. In the first study[48] the fats compared were elaidic acid, oleic acid, and corn oil. Elaidic acid was significantly more cholesterolemic than oleic acid and both were significantly hypercholesterolemic compared to corn oil. However, there was no significant difference between the percentage of aortic lesions due to either monoenic acid. In a second study[49] the fats used were elaidinized olive oil and olive oil. As in the earlier experiments the *trans* fat was significantly more cholesterolemic but not atherogenic. The third study[50] involved comparison of linolelaidic and linoleic acids and a fat-free group was also included. In this study the *trans* fat was only slightly more cholesterolemic than the *cis* fat but not significantly more atherogenic. The fat-free group exhibited a lower cholesterol level than either of the two fat-fed groups but the diet was almost as atherogenic as that of the group fed *trans* fat.

Vles *et al.*[51] carried out a 6 year study in which rabbits were fed soybean or coconut oil or soybean oil hydrogenated to three different levels. Both

Table 5 *Influence of trans fat (6%) on atherosclerosis in rabbits fed 1% cholesterol*

Fats fed	Serum cholesterol (mmol/L)	Aortic lesions (%)
Experiment 1[a]		
Elaidic acid	99.0 ± 6.3	37 ± 5
Oleic acid	62.4 ± 5.8	32 ± 5
Corn oil	21.7 ± 3.9	10 ± 3
Experiment 2[b]		
Elaidimized olive oil	63.3 ± 6.4	30 ± 4
Olive oil	37.2 ± 4.2	23 ± 5
Experiment 3[c]		
Linolelaidic acid	39.5 ± 4.2	33 ± 8
Linoleic acid	36.5 ± 5.2	21 ± 6
Fat free	25.2 ± 3.4	28 ± 7

[a] After Weigensberg et al.[48]
[b] After McMillan et al.[49]
[c] After Weigensberg and McMillan.[50]

mortality over the 6 years and severity of atherosclerosis were related more to level of saturation than to percentage of *trans* fatty acids, the greatest number of lesions appearing in the rabbits fed coconut oil. Gottenbos[52] fed rabbits mixtures of saturated, unsaturated, or hydrogenated oils for 2 years. There were no differences in atherogenicity between saturated or *trans* fat. One study was carried out in rabbits fed semipurified, cholesterol-free, atherogenic diets[53] containing 3.2 or 6.0% 18:1*t* or a mixture of unsaturated fats. The *trans* fat-rich diets were more cholesterolemic than the control fat but not more atherogenic (Table 6).[37]

Table 6 *Influence of trans fat on atherosclerosis in rabbits fed semipurified, cholesterol-free diets*[a]

		Regimen	
	3.2% 18:1*t*	6.0% 18:1*t*	Control
Plasma lipids (mm/L)			
Cholesterol	2.25 ± 0.47	3.54 ± 1.34	1.86 ± 0.26
Triglycerides	0.78 ± 0.01	0.78 ± 0.01	0.84 ± 0.02
Average atherosclerosis			
Aortic arch	0.2 ± 0.1	0.2 ± 0.1	0.2 ± 0.1
Thoracic aorta	0.1 ± 0.1	0	0

[a] After Ruttenberg et al.[37] Diets contained 14% fat. Aortas graded on a 0–4 scale.

Kummerow[54] reported that a diet containing *trans* fat was more atherogenic for swine than a basal diet but the results may have been due to EFA deficiency in the test diet. A second experiment in which *trans* fats were fed in an EFA-replete diet also showed no differences in atherogenicity between the *trans* fat and the other diets.[17] A third study[55] comparing swine fed *trans* or *cis* fat in an EFA-replete diet showed no differences in atherogenicity between groups.

Elson et al.[56] fed young swine fat blends varying in *trans* fat content from 0 to 48%. The level of fat in the diet was 17%. After 10 months the levels of plasma cholesterol in the groups fell within a 10% range and only 5 of 64 pigs exhibited aortic sudanophilia that covered more than 2% of the aortic surfaces. The five pigs came from groups fed 0, 24.8, or 48% *trans* fat.

Monkeys were fed a control diet or the same diet containing 3.2 or 6.0% *trans* fat for a year. Two other groups were fed the *trans* fat-rich diets for 6 months then placed on the control diet for 6 months. There were no significant differences between cholesterolemia or atherosclerosis[24] (Table 7). Tissue levels of *trans* fat reflected the concentration present in the diet. Six months after reversion to the control diet *trans* fat had virtually disappeared from the plasma and was greatly reduced in levels of the monkeys fed 3.2 or 6.0% *trans* fat.

Table 7 *Influence of trans fat on atherosclerosis in monkeys*[a]

Regimen[b]	Cholesterol (mm/L)	Atherosclerosis Incidence	% of surface
3T	3.47 ± 0.31	6/8	2.5 ± 1.3
3R	4.34 ± 0.44	6/8	5.1 ± 2.6
6T	4.22 ± 0.34	5/8	5.3 ± 4.0
6R	3.78 ± 0.13	6/7	3.0 ± 1.0
Control	4.29 ± 0.21	12/15	6.6 ± 2.4

[a] After Kritchevsky et al.[24]
[b] 3T = fed 3.2% 18:1*t* for 12 months; 3R = fed 3.2% 18:1*t* for 6 months, then control fat (soybean oil) for 6 months; 6T = fed 6.0% 18:1*t* for 12 months; 6R = fed 6.0% 18:1*t* for 6 months, then control fat for 6 months.

Plasma Cholesterol in Man

Concern about the effects of *trans* fats on human plasma cholesterol levels dates to the 1960s. The concern is due to the role of plasma cholesterol as a risk factor for coronary heart disease. The risk factors, which are based on epidemiological observations, are useful on a population basis but much less precise indicators for individuals. McOsker et al.[57] fed male volunteers seven different fats ranging in iodine number (degree of unsaturation) from 32 to 114. Four of the fats were partially hydrogenated, the most unsaturated fat was cottonseed oil, and the most saturated was butterfat. There were four groups of six subjects each and they were fed diets containing the various fats in rotation. The four hydrogenated fats contained 14–21% *trans* fatty acids. The fats with the lowest content of saturated fatty acids gave the lowest cholesterol levels regardless of *trans* fatty acid content. In another study Erickson et al.[58] fed seven different groups of six men each a number of diets containing varying levels of saturated, unsaturated, or *trans* fatty acids. There were no differences in cholesterol level among the groups. Addition of cholesterol to the diets raised the probands' cholesterol levels but the increase was the same whether or not the diet contained *trans* fat. In 1975, Mattson *et* al.[59] carried out a similar experiment and again found no effect of *trans* fat on plasma cholesterol levels.

The role of *trans* fats in cholesterolemia was re-investigated by Mensink and Katan[60] 15 years later. They fed a group of male and female volunteers diets rich in oleic acid, isomerized oleic acid (yielding mostly elaidic acid), or saturated fat. The fats represented about 40% of total calories. Iodine values (calculated) for the three fats were oleic 76, *trans* fat 85, saturated fat 59. Cholesterol levels (mm/L) and LDL-cholesterol/HDL-cholesterol in subjects ingesting the three diets were: 4.45±0.67 and 1.88; 4.81±0.72 and 2.43; and 4.99±0.70 and 2.21. All differences were significant. The oleic acid-rich diet also gave lowest triglyceride levels. Nestel et al.[61] studied 27 men who were taken from their habitual diet or the same diet altered by adding an oleic acid-rich fat, elaidic acid-rich fat, or palmitic acid-rich fat. The elaidic acid was added to provide twice the usual *trans* fat content of the Australian diet. The oleic acid addition lowered the plasma cholesterol level compared to the habitual diet (5.56 mm/L vs. 5.90 mm/L or 5.7%) and the LDL/HDL cholesterol ratio fell by 7.5%. The plasma cholesterol levels in the other two groups (elaidic and palmitic) were unchanged from that observed in the habitual diet. The addition of elaidic acid raised the LDL/HDL cholesterol level by 1.2% and the palmitic acid diet lowered it by 10.7%. Zock and Katan[62] fed 56 volunteers (male and female) diets rich in linoleate, stearate, or *trans* fat in varying schedules. The regimen was 3 week periods of linoleate-stearate-*trans*, linoleate-*trans*-stearate, and similar variations in which the initial feeding period contained stearate or *trans*. The fatty acids of the linoleate diet contained 18.8% mono- and 62.5% polyunsaturated fat and 7% stearate; the stearate diet contained 45.5% stearate, 37.4% mono- and 9.5% polyunsaturated fat; and the *trans* diet contained 6.6% stearate, 2.1% polyunsaturated fat, 43.6% cis monounsaturates, and 40.2% *trans* monounsaturates. It should be pointed out that the level of *trans* fat was abnormally high and that stearic acid does not exert the usual hypercholesterolemic effect attributed to saturated fatty acids such as myristic or palmitic. The linoleate diet led to cholesterol levels which were 3% lower than those seen in the other two groups, which exhibited the same levels of cholesterol. The HDL/LDL cholesterol ratio was reduced by 9.1% (compared to linoleate) on the stearate diet and by 14.5% on the *trans* fat diet. The finding that the *trans* fat reduced HDL cholesterol levels (and hence increased risk) was exciting.

Lichtenstein et al.[63] took 14 subjects from a baseline diet and fed them on diets rich in corn oil or margarine (4.2% 18:1$n-9$ *trans*). After 32 days both test diets had lowered cholesterol levels, the corn oil diet by 13.9% and the margarine diet by 8.1%. The LDL/HDL was lowered by 11% on the corn oil diet and by 2% on the margarine. There was no effect of the margarine on Lp(a) levels although Mensink et al.[64] had reported that *trans* fat increased Lp(a) levels. Wood et al.[65] compared effects of diets containing butter, butter plus sunflower oil, butter plus olive oil, soft margarine, and hard margarine with a baseline diet in 38 male subjects. Total cholesterol levels were raised by 4.5% on butter and reduced by 7.4 and 2.0% by soft and hard margarine, respectively. The serum HDL cholesterol levels were the same in all six groups (46±0.40 mg/dL).

Judd et al.[66] studied a group of 29 men and 29 women who ate each of the following diets for 6 weeks in a Latin square design: (a) high oleic acid, 16.7% of energy; (b) moderate *trans* fatty acids, 3.8% of energy; (c) high *trans* fatty acids, 6.6% of energy; and (d) 16.2% of energy as saturated fat (lauric, myristic, and palmitic acids). Plasma cholesterol levels (mm/L) and LDL/HDL cholesterol ratios were: diet (a) 5.25±0.26 and 1.18; (b) 5.46±0.26 and 1.26; (c) 5.51±0.26 and 1.30; and (d) 5.61±0.26 and 1.24. The starting levels were 5.30±0.59 and 1.22. Thus, cholesterol levels were unchanged on the oleic acid-rich diet, elevated by 3 and 4% on the two *trans* fat diets and 6% on the saturated fat diet. As in many of the other studies, saturated fat was more cholesterolemic than *trans* fat. Plasma Lp(a) levels were not affected by the diets.[67] Lp(a) is a variant of apolipoprotein B which carries an additional peptide chain. Lp(a) interferes with fibrinolysis and poses a risk because blood clots persist in the circulation.

It is evident that the effects of the dietary interventions must be judged against the background of the habitual diet. Aro et al.[68] placed subjects (40 per group) on a dairy-fat rich diet then fed diets containing stearic acid (9.3% of energy) or *trans* fat (8.7% of energy). Both stearic acid and *trans* fat *lowered* serum total cholesterol but both lowered HDL-cholesterol levels. Both experimental diets increased Lp(a) levels. It should be recognized that the levels of stearic acid and *trans* fat (about 9% of energy) were considerably higher than might be consumed in a regular daily diet.

Table 8 *Influence of trans fat and linoleic acid on serum cholesterol levels in man*[a]

Fatty acid (%) Trans (A)	18:2 (B)	Change in A/B	cholesterol (%)[b]	Ref.
35	13	2.69	+25	56
27	11	2.45	+21	56
18	6	3.00	+19	70
27	11	2.45	+10	60
19	10	1.90	+6	62
18	22	0.81	+4	57
10	37	0.27	+1	69
11	31	0.35	+4	58
8	33	0.24	+3	71

[a] After Emken.[69]
[b] $r = 0.75$.

Overall, the pattern that emerges is that *trans* fats are hypercholesterolemic compared to *cis* fats but not as hypercholesterolemic as are saturated fats. *Trans* fats may reduce HDL-cholesterol levels to a slight degree and effects on Lp(a) levels are inconsistent. Emken[69] approached this dichotomy as being due to the ratio of *trans* fatty acids to linoleic acid. His summary of the data (Table

8) includes papers not discussed above. It is evident that the presence of enough dietary linoleic acid modulates the effects of *trans* fat.

Tissue Levels

A better test of *trans* fat effects might be assessment of their presence in body fat stores. Heckers et al.[72] determined levels of 14:1t, 16:1t, and 18:1t in myocardium, aorta and jejunum of subjects with advanced atherosclerosis and found them to be in the same ranges as those of controls (Table 9). Aortic 18:1t was 58% higher in the controls and myocardial 18:1t was 7% higher. Thomas et al.[73] measured adipose tissue *trans* fatty acid levels of coronary patients and controls in 10 areas of England. There were 136 samples from coronary patients and 95 samples from controls. Average levels of *trans* fat were 5.25% and 5.38%, respectively. Roberts et al.[74] analysed the fatty acids of adipose tissue obtained from 66 cases of sudden cardiac death and compared the results with those obtained upon analysis of adipose tissue of 286 age and sex matched controls. The total of *trans* fatty acids in the cases (2.68±0.08%) was significantly lower than in the healthy controls (2.86±0.04%) ($p<0.05$). They concluded, 'overall, there was no evidence of a relation between *trans* isomers of oleic and linoleic acids combined and sudden cardiac death'. Aro et al.[75] addressed the issue of tissue *trans* fat and risk of myocardial infarction in nine countries (eight European countries and Israel). They compared data from 671 men with acute myocardial infarction and 717 controls. There was no difference between adipose *trans* fatty acids in cases (1.61±0.92%) and controls (1.57±0.86%).

Table 9 *Tissue trans fatty acids (%) in subjects with advanced atherosclerosis and control subjects*[a]

Tissue	No.		Fatty acid	
		14:1t	16:1t	18:1t
Myocardium				
Atherosclerosis	15	0.05 ± 0.02	0.36 ± 0.12	0.41 ± 0.31
Controls	8	0.06 ± 0.02	0.48 ± 0.08	0.44 ± 0.29
Aorta				
Atherosclerosis	10	0.06 ± 0.02	0.61 ± 0.20	0.33 ± 0.12
Controls	7	0.05 ± 0.02	0.56 ± 0.06	0.52 ± 0.29

[a] After Heckers et al.[72]

Hudgins et al.[76] studied the adipose tissue concentrations of 19 geometrical and positional fatty acid isomers and 10 cardiovascular risk factors in 76 adult male subjects. Total *trans* isomers and most individual isomers were not related significantly to risk factors for atherosclerosis. They found no significant relationships between isomeric fatty acids and plasma triglycerides, HDL-cholesterol, LDL/HDL cholesterol, or total/HDL cholesterol. Lauric, myristic, and stearic acids were negatively associated with plasma lipids. There was a tendency of 20:3$n-6$, 20:4$n-6$, and 22:5$n-3$ to be associated positively with plasma total and LDL-cholesterol, LDL/HDL-cholesterol, and total/HDL-cholesterol. They also found positive relations between 20:4$n-6$ and LDL-cholesterol, diastolic blood pressure, and body mass index. van de Vijver et al.[77] studied the relationship between *trans* fatty acids in phospholipids and coronary heart disease. The study included 83 cases with angiographically documented severe coronary heart disease and 78 controls with minor stenosis. They found no correlations between the percentage of *trans* fatty acids in plasma phospholipids and plasma LDL or HDL cholesterol levels. Their conclusion was that the findings did not support an association between the intake of *trans* fat and risk of coronary heart disease.

Epidemiology

In an epidemiological study Willett et al.[78] used dietary questionnaires to calculate the *trans* fat intake of 85 095 women who were free of heart disease, diabetes, or stroke. Over an 8 year period there were 431 (0.51%) new cases of heart disease. Comparing lowest to highest quintile of intake there was a significant positive risk for coronary heart disease. The trend was significant for total *trans* fat or *trans* fat of vegetable origin but not *trans* fat of animal origin. The relative risk (by quintile) as a function of energy-adjusted *trans* fat intake was 1.00, 1.12, 0.99, 1.16, and 1.47. It would appear that the major problem was in the fifth quintile of intake (5.7 g/day) which is below the national average. Dietary fiber intake was highest in the first quintile (19.3 g/day) and fell in each succeeding quintile to 14.0 g/day in the fifth quintile. Ascherio et al.[79] compared *trans* fat intake of 239 subjects who had suffered an acute myocardial infarction with 282 population controls. They reported a significant trend towards increased risk in going from the first to the fifth quintile of *trans* fat intake. When adjusted for age, sex, smoking, hypertension, body mass index, alcohol intake, family history of heart disease, and physical activity, relative risk by quintile was: 1.0, 0.89, 0.52, 0.93, 2.28. When further adjusted for intake of saturated and monounsaturated fat, linoleic acid, and cholesterol intake, the relative risk by quintile became: 1.0, 0.81, 0.40, 0.72, and 2.03. As in the earlier study[78] there appeared to be no risk attached to *trans* fat from animal sources. Relative to the last quintile of intake, one wonders how many of the subjects exhibited risk factors for coronary disease and were advised to substitute margarine for butter.

2 Discussion

The foregoing discussion suggests that one can adduce a risk factor relationship between *trans* fat and heart disease but the data do not lean towards unanimity. Concern about deleterious effects of dietary *trans* fats is not new. More than a decade ago there was pressure on the medical and nutritional communities to address this problem.

Expert committees of the Life Sciences Research Office of the Federation of American Societies of Experimental Biology[80] and the British Nutrition

Foundation[81] independently concluded that the level of intake of *trans* fat posed no measurable risk. Ten years later a new report from the British Nutrition Foundation[82] and one issued by an expert committee organized by the International Life Sciences Institute[83] came to the conclusion reached earlier, namely, that *trans* fat at current levels of intake poses no health threat. This conclusion was echoed by a point task force from the American Institute of Nutrition and the American Society for Clinical Nutrition.[84] All of the reports stressed the need for more research.

More research is indeed indicated. Wahle and James[85] have raised questions regarding *trans* fat effects on the human fetus and the neonate. This concern has also been voiced by a committee of the Danish Nutrition Council.[86] More has to be learned about specific effects of specific *cis* and *trans* unsaturated fatty acids. This is especially true considering the array of these fatty acids that arises during hydrogenation. The current literature provides conflicting data regarding effects of *trans* fat on plasmas Lp(a) and HDL-cholesterol levels. Are the disparate findings due to specific fatty acids or due to interactions with other dietary components such as type of protein or fiber or to both? Could there be effects of specific *trans* fats on gene expression? All of these possibilities should be investigated. At the moment, the suggestion of Houtsmüller[15] that *trans* fats be regarded physiologically as saturated fats merits serious consideration. *Trans* fats represent a small fraction of our fat intake and while they may exert some untoward properties, they should not deflect our attention from the broader aspects of fat and fatty acid effects on health which, it should be noted, are not all deleterious.

Table 10 *Trans fat in the American diet and mortality*

| Year | Trans fat[b] (g/person/day) | Mortality (deaths/100,000)[a] | | Heart disease | Stroke |
		Crude	Age adjusted		
1970	7.70	945.3	714.3	253.6	66.3
1975	7.83				
1980	7.55	878.3	585.8	202.0	40.8
1984	7.72				
1985	–	873.9	546.1	180.5	32.3

[a] Health United States, 1994.[88]
[b] After Hunter and Applewhite.[87]

In the United States, availability of *trans* fat has been fairly constant since 1963 (about 7.5 g/person/day)[87] and while all-cause, age-adjusted mortality fell by 7.6% between 1970 and 1985, death from heart disease fell by 28.8% and death from stroke fell by 51.3% (Table 10).[88] The data do not support the view of *trans* fat as a killer. We have learned a lot about the biological effects of *trans* fat. A book has recently been published which deals with this subject.[89] We need to learn more and we are working towards that end. In the meantime, the best suggestion vis-a-vis diet is moderation, variety, and balance. Prudence not panic!

References

1. M. Sommerfeld, *Prog. Lipid Res.*, 1983, **22**, 221.
2. J. Grandgirard, J. L. Sebedio and J. Fleury, *J. Am. Oil Chem. Soc.*, 1984, **61**, 1563.
3. J. L. Sebedio, A. Grandgirard and J. Prevost, *J. Am. Oil Chem. Soc.*, 1988, **65**, 362.
4. A. Kiritsakis, F. Aspris and P. Markakis, *Flavors and Off-Flavors*, Elsevier, Amsterdam, 1989, p. 893.
5. R. G. Ackman, S. N. Hooper and D. L. Hooper, *J. Am. Oil Chem. Soc.*, 1979, **51**, 42.
6. H. J. Dutton, *Geometrical and Positional Fatty Acid Isomers*, Am. Oil Chem. Soc., Champaign, IL, 1979, p. 1.
7. D. Melnick and H. J. Deuel Jr., *J. Am. Oil Chem. Soc.*, 1954, **31**, 63.
8. F. H. Mattson, *J. Nutr.*, 1960, **71**, 366.
9. E. Aaes-Jorgenson, J. P. Funch and H. Dam, *Br. J. Nutr.*, 1956, **10**, 317.
10. R. B. Alfin-Slater, P. Wells, L. Aftergood and D. Melnick, *J. Am. Oil Chem. Soc.*, 1973, **50**, 479.
11. R. B. Alfin-Slater and L. Aftergood, *Geometrical and Positional Fatty Acid Isomers*, Am. Oil Chem. Soc. Champaign, IL, 1979, p. 53.
12. I. F. Duthie, S. H. Barlow, R. Ashby, J. M. Tesh, J. C. Whitney, A. Saunders, E. Chapman, K. R. Norum, H. Svaar and J. Opstvedt, *Acta Med. Scand.*, 1988, Suppl. 726.
13. J. L. Zevenbergen, U. M. T. Houtsmüller and J. J. Gottenbos, *Lipids*, 1988, **23**, 178.
14. J. J. Gottenbos and H. J. Thomasson, *Bibl. Nutr. Dieta*, 1965, **7**, 110.
15. U. M. J. Houtsmüller, *Fette, Seifern, Anstrichm.*, 1978, **80**, 162.
16. A. J. Vergroesen, *Proc. Nutr. Soc.*, 1972, **31**, 323.
17. R. L. Jackson, J. D. Morrisett, H. J. Pownall, A. M. Gotto Jr., A. Kamio, H. Imar, R. Tracy and F. A. Kummerow, *J. Lipid Res.*, 1977, **18**, 182.
18. M. Sugano, M. Watanabe, M. Kohno, Y. J. Cho and T. Ide, *Lipids*, 1983, **18**, 375.
19. R. B. Sinclair, *J. Biol. Chem.*, 1935, **111**, 515.
20. W. J. Decker and W. Mertz, *J. Nutr.*, 1966, **89**, 165.
21. D. Sgoutas and F. Kummerow, *Am. J. Clin. Nutr.*, 1970, **23**, 1111.
22. R. Wood, *Lipids*, 1979, **14**, 975.
23. C. E. Moore, R. B. Alfin-Slater and L. Aftergood, *Am. J. Clin. Nutr.*, 1980, **33**, 2318.
24. D. Kritchevsky, L. M. Davidson, M. Weight, N. P. J. Kriek and J. P. duPlessis, *Atherosclerosis*, 1984, **51**, 123.
25. K. Ono and D. S. Fredrickson, *J. Biol. Chem.* 1964, **239**, 2482.
26. R. H. Coots, *J. Lipid Res.*, 1964, **5**, 468.
27. R. H. Coots, *J. Lipid Res.*, 1964, **5**, 473.
28. L. D. Lawson and F. A. Kummerow, *Lipids*, 1979, **14**, 501.
29. L. D. Lawson and F. A. Kummerow, *Biochim. Biophys. Acta*, 1979, **573**, 245.
30. D. S. Privett, E. M. Stearns Jr. and E. C. Nickell, *J. Nutr.*, 1967, **92**, 303.
31. P. O. Egwim and D. S. Sgoutas, *J. Lipid Res.*, 1971, **13**, 307.
32. P. O. Egwim and F. A. Kummerow, *J. Lipid Res.*, 1972, **13**, 500.
33. T. Takatori, F. C. Phillips, H. Shemasaki and O. S. Privett, *Lipids*, 1976, **11**, 272.
34. C. E. Moore, R. B. Alfin-Slater and L. Aftergood, *J. Nutr.*, 1980, **110**, 2284.

35. D. Kritchevsky and A. R. Baldino, *Artery*, 1978, **4**, 480.
36. D. S. Sgoutas, *Biochim. Biophys. Acta*, 1968, **164**, 317.
37. H. Ruttenberg, L. M. Davidson, N. A. Little, D. M. Klurfeld and D. Kritchevsky, *J. Nutr.*, 1983, **113**, 835.
38. T. Koga, T. Yamato, J. Gu, M. Monaka, K. Yamada and M. Sugano, *Biosci. Biotechnol. Biochem.*, 1994, **58**, 384.
39. R. R. Brown, *Cancer Res.*, 1981, **41**, 3741.
40. S. L. Selenskas, M. M. Ip and C. Ip, *Cancer Res.*, 1984, **44**, 1321.
41. D. Kritchevsky, M. M. Weber and D. M. Klurfeld, *Nutr. Res.*, 1992, **12**, S175.
42. K. L. Erickson, D. S. Schlanger, D. A. Adams, D. R. Fregan and J. S. Stern, *J. Nutr.*, 1984, **114**, 1834.
43. B. S. Reddy, T. Tanaka and B. Simi, *J. Natl. Cancer Inst.*, 1985, **75**, 791.
44. M. Watanabe, T. Koga and M. Sugano, *Am. J. Clin. Nutr.*, 1985, **42**, 475.
45. S. Sugano, M. Watanabe, K. Yoshida, M. Tomioka, M. Miyamoto and D. Kritchevsky, *Nutr. Cancer*, 1989, **12**, 177.
46. C. Ip, C. A. Carter and M. M. Ip, *Cancer Res.*, 1977, **45**, 1997.
47. K. K. Carroll and T. H. Khor, *Lipids*, 1971, **6**, 415.
48. B. I. Weigensberg, G. C. McMillan and A. C. Ritchie, *Arch. Pathol.*, 1961, **72**, 126.
49. G. C. McMillan, M. D. Silver and B. I. Weigensberg, *Arch. Pathol.*, 1963, **76**, 106.
50. B. I. Weigensberg and G. C. McMillan, *Exp. Mol. Pathol.*, 1964, **3**, 201.
51. R. O. Vles, J. J. Gottenbos and P. L. Van Pijpen, *Bibl. Nutr. Dieta*, 1977, **25**, 186.
52. J. J. Gottenbos, *Dietary Fats and Health*, Am. Oil Chemists Soc., Champaign, IL, 1983, p. 375.
53. D. Kritchevsky and S. A. Tepper, *J. Atheroscler. Res.*, 1968, **8**, 357.
54. F. A. Kummerow, *J. Food Sci.*, 1975, **40**, 12.
55. S. M. Royce, R. P. Holmes, T. Takagi and F. A. Kummerow, *Am. J. Clin. Nutr.*, 1984, **39**, 215.
56. C. E. Elson, N. J. Benevenga, D. J. Canty, R. H. Grummer, J. J. Talich, J. W. Porter and A. E. Johnston, *Atherosclerosis*, 1981, **40**, 115.
57. D. E. McOsker, F. H. Mattson, H. B. Sweringen and A. M. Kligman, *J. Am. Med. Assoc.*, 1962, **180**, 380.
58. B. A. Erickson, R. H. Coots, F. H. Mattson and A. M. Kligman, *J. Clin. Invest.*, 1964, **43**, 2017.
59. F. H. Mattson, E. J. Hollenbach and A. M. Kligman, *Am. J. Clin. Nutr.*, 1975, **28**, 726.
60. R. P. Mensink and M. B. Katan, *N. Engl. J. Med.*, 1990, **323**, 439.
61. P. Nestel, M. Noakes, B. Belling, R. McArthur, P. Clifton, E. James and M. Abbey, *J. Lipid Res.*, 1992, **33**, 1029.
62. P. L. Zock and M. B. Katan, *J. Lipid Res.*, 1992, **33**, 399.
63. A. H. Lichtenstein, L. M. Ausman, W. Carrasco, J. L. Jenner, J. M. Ordovas and E. J. Schaefer, *Arterioscler. Thromb.*, 1993, **13**, 154.
64. R. P. Mensink, P. L. Zock, M. B. Katan and G. Hornstra, *J. Lipid Res.*, 1992, **33**, 1493.
65. R. Wood, K. Kubena, B. O'Brien, S. Tseng and G. Martin, *J. Lipid Res.*, 1993, **34**, 1.
66. J. T. Judd, B. A. Clevidence, R. A. Muesing, J. Wittes, M. E. Sunkin and J. J. Podczasy, *Am. J. Clin. Nutr.*, 1994, **59**, 861.

67. B. A. Clevidence, J. T. Judd, E. J. Schaefer, J. R. McNamara, R. A. Muesing, J. Wittes and M. E. Sunken, *Arterioscler. Thromb. Vasc. Biol.*, 1997, **17**, 1657.
68. A. Aro, M. Jaukiainen, R. Partanen, I. Salminen and M. Mutanen, *Am. J. Clin. Nutr.*, 1997, **65**, 1419.
69. E. A. Emken, *Fat Nutr.*, 1992, **1** (2), 1.
70. H. DeIongh, R. K. Beerthuis, C. den Hartog, L. M. Dalderup and P. A. F. van der Spek, *Bibl. Nutr. Dieta*, 1965, **7**, 137.
71. D. C. Laine, C. M. Snodgrass, E. A. Dawson, M. A. Ener, K. Kuba and I. D. Frantz, *Am. J. Clin. Nutr.*, 1982, **35**, 683.
72. H. Heckers, M. Korner, T. W. L. Tuschen and F. W. Melcher, *Atherosclerosis*, 1977, **28**, 389.
73. L. H. Thomas, P. R. Jones, J. A. Winter and H. Smith, *Am. J. Clin. Nutr.*, 1981, **34**, 877.
74. T. L. Roberts, D. A. Wood, R. A. Riemersma, P. J. Gallagher and F. C. Lampe, *Lancet*, 1995, **345**, 278.
75. A. Aro, A. F. M. Kardinaal, I. Salminen, J. D. Kark, R. A. Riemersma, M. Delgado-Rodriguez, J. Gomez-Aracena, J. K. Huttunen, L. Kohlmeier, B. C. Martin, J. M. Martin-Moreno, V. P. Mazaev, J. Ringstad, M. Thamm, P. Van't Veer and F. J. Kok, *Lancet*, 1995, **345**, 273.
76. L. C. Hudgins, J. Hirsch and E. A. Emken, *Am. J. Clin. Nutr.*, 1991, **53**, 474.
77. L. P. L. van de Vijver, G. van Poppel, A. van Houwelingen, D. A. C. M. Kruyssen and G. Hornstra, *Atherosclerosis*, 1996, **126**, 155.
78. W. C. Willett, M. J. Stampfer, J. E. Manson, G. A. Colditz, F. E. Speizer, B. A. Rosner, L. A. Sampson and C. H. Hennekens, *Lancet*, 1993, **341**, 581.
79. A. Ascherio, C. H. Hennekens, J. E. Biering, C. Master, M. J. Stampfer and W. C. Willett, *Circulation*, 1994, **89**, 94.
80. F. R. Senti (ed.) *Health Aspects of Dietary Trans Fatty Acids*, FASEB, Bethesda, MD, 1985.
81. British Nutrition Foundation, *Trans Fatty Acids*, Br. Nutr. Foundation, London, 1987.
82. British Nutrition Foundation, *Trans Fatty Acids*, Br. Nutr. Foundation, London, 1995.
83. P. K. Etherton (ed.) *Am. J. Clin. Nutr.*, 1995, **62**, 655S.
84. ASCN/AIN Task Force on Fatty Acids, *Am. J. Clin. Nutr.*, 1996, **63**, 463.
85. K. W. J. Wahle and W. P. T. James, *Eur. J. Clin. Nutr.*, 1993, **47**, 828.
86. S. Stender, J. Dyerberg, G. Holmer, I. Oveson and B. Sandstrom, *Clin. Sci.* 1995, **88**, 375.
87. J. E. Hunter and T. H. Applewhite, *Am. J. Clin. Nutr.*, 1991, **54**, 363.
88. National Center for Health Statistics, *Health United States 1994*, Hyattsville, MD, 1995.
89. J. L. Sebedo and W. W. Christie, *Trans Fatty Acids in Human Nutrition*, The Oily Press, Dundee, Scotland, 1998.

4
Antioxidant Properties of Flavonols

Michael H.Gordon and Andrea Roedig-Penman

DEPARTMENT OF FOOD SCIENCE AND TECHNOLOGY,
UNIVERSITY OF READING, PO BOX 226, WHITEKNIGHTS,
READING RG6 6AP, UK

1 Autoxidation

Lipids occur widely in biological systems either as structural lipids including phospholipids, sphingolipids and glycolipids or as triacylglycerols which are the main lipids involved as energy stores. Unsaturated fatty acyl groups in lipids may react with oxygen by a reaction known as autoxidation. This reaction leads to the development of off-flavours in foods, commonly termed rancidity, and it also occurs *in vivo* and leads to chronic health problems including coronary heart disease, cancer and rheumatoid arthritis if the body's defence systems are overwhelmed by the radicals generated.

Oxidative deterioration of lipids can occur by two possible routes: either by the classical free radical mechanism, known as autoxidation, which can occur in the dark, or by a photo-oxidation mechanism, which is initiated by exposure to light. The intermediate in both cases is a lipid hydroperoxide. These are colourless and odourless compounds which decompose easily to form secondary oxidation products which are responsible for the off-flavours in foods. Different fatty acids vary in their oxidation rates with polyunsaturated fatty acid: *e.g.* linoleic and linolenic acids oxidise much more rapidly than monounsaturated fatty acids, *e.g.* oleic acid.

Classical Free Radical Route

The classical free radical route consists of three steps: initiation, propagation and termination. The initiation step consists of the production of a lipid free radical. Commonly, decomposition of hydroperoxides is the main pathway to lipid radicals, since hydroperoxides are usually present owing to the reaction of the enzyme lipoxygenase with a polyunsaturated fatty acid, or due to the reaction of singlet oxygen with an unsaturated fatty acid. However, R$^{\cdot}$ can be formed from a lipid molecule RH by interaction with oxygen in the presence of a catalyst, *e.g.* the initiation can occur by the action of external energy sources such as light, heat, high energy irradiation or by chemical initiation involving metal ions or metalloproteins such as haem. Non-lipid radicals can also react with lipid molecules to form a lipid radical. The free radical R˙ produced in the initiation step can react further to form a lipid peroxy radical ROO˙ which can then react further to give the hydroperoxide ROOH. The propagation step also provides a further free radical R˙ making it a self-propagating chain process. Only a small amount of catalyst is required to cause the formation and breakdown of many hydroperoxide molecules. The propagation can be stopped by a termination reaction, where two radicals combine to give products which do not feed the propagating reactions, but although having a low energy of activation termination reactions are limited by the low probability of two radicals colliding with the correct orientation, if the radical concentration is low. A simple scheme for autoxidation can be summarised as in Figure 1.

$$\text{Initiation:} \quad RH + O_2 \xrightarrow{cat.} R\cdot + \cdot OOH$$

$$RH \xrightarrow{cat.} R\cdot + H\cdot$$

$$\text{Propagation:} \quad R\cdot + O_2 \longrightarrow RO_2\cdot$$

$$RO_2\cdot + RH \longrightarrow RO_2H + R\cdot$$

$$\text{Termination:} \quad R\cdot + R\cdot \longrightarrow R\text{-}R$$

$$RO_2\cdot + R\cdot \longrightarrow RO_2R$$

RH: unsaturated lipid; R˙: lipid radical; $RO_2\cdot$: lipid peroxy radical

Figure 1 *The mechanism of autoxidation*

When the fatty acid hydroperoxide is formed in the propagation step, a migration of the double bond occurs in the case of polyunsaturated fatty acids to form a conjugated system.

Secondary Oxidation Products

As lipid peroxides are very unstable they break down to an alkoxy free radical which decomposes *via* cleavage of the bond to the carbon atom bearing the oxygen atom to form a number of secondary oxidation products such as alcohols, aldehydes, alkanes, ketones, etc. The number of volatile products arising from oxidation of an unsaturated fatty acid is so large because each of the primary hydroperoxides is a precursor for several volatile compounds.[1]

Linoleic acid is the major polyunsaturated fatty acid in vegetable oils and biological tissues. Free radical autoxidation of this fatty acid produces a mixture of *cis, trans*- and *trans,trans*-conjugated diene 9- and 13-hydroperoxides. The major volatile carbonyls formed from linoleic acid include hexanal, 2,4-decadienal and methyl 9-oxononanoate. Hexanal is formed as the major volatile at milder temperatures, with 2,4-decadienal being formed at significant concentrations at higher temperatures. A series of C_1 to C_5 hydrocarbons are formed with pentane being the predominant one.

Induction Period

The oxidation of lipids proceeds through two different phases (Figure 2). During the first phase, called the induction period, the oxidation goes slowly and at a uniform rate. After the oxidation has proceeded to a certain point the reaction enters a second phase, which has a rapidly accelerating rate of oxidation. Even before the second phase is reached one can notice off-flavours in edible oils, but they are particularly noticeable from the beginning of the second phase.

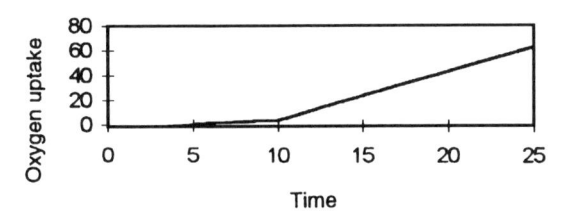

Figure 2 *Change in oxygen uptake with time demonstrating the presence of the induction period*

Metal-catalysed Lipid Oxidation

Transition metals, *e.g.* Fe, Cu, Co, which possess two or more valence states with a suitable oxidation–reduction potential play a key role in lipid oxidation in biological systems, affecting both the speed of autoxidation and the direction of hydroperoxide breakdown to volatile compounds.[1]

Transition metal ions in their lower valence state (M^{n+}) react very fast with hydroperoxides. They act as one-electron donors to form an alkoxy radical; this can be considered as the branching of the propagation step:

$$ROOH + M^{n+} \rightarrow RO^{\cdot} + OH^{-} + M^{(n+1)+} \qquad (1)$$

In a slow consecutive reaction the reduced state of the metal ion may always be regenerated by hydroperoxide molecules:

$$ROOH + M^{(n+1)+} \rightarrow ROO^{\cdot} + H^{+} + M^{n+} \qquad (2)$$

Owing to the presence of water or the metal complexing with chain termination products, the regeneration is not complete. The radicals produced in both the above equations enter the propagation sequence and decrease the induction period. The catalytic activity of heavy metals depends in reality not only on the ion species and its redox potential but also on the ligands attached to it, the solvent system, the pH and the presence of electron donors such as ascorbate and cysteine which keep the metal ion in its lower valence state. Maximum degradation of peroxides occurs in pH region of 5–5.5 and the activity decreases from $Fe^{2+} > Fe^{3+} > Cu^{2+}$.[2]

It is not clear if heavy metals can abstract a hydrogen atom from the fatty acids themselves:

$$RH + M^{(n-1)+} \rightarrow R^{\bullet} + H^+ + M^{n+} \qquad (3)$$

However, this reaction would be slower than hydrogen abstraction by a lipid hydroperoxide radical.[3]

Catalysis by Haem Compounds

Haem (Fe^{2+}) and haemin (Fe^{3+}) compounds are widely distributed in plants and animals. Haemoglobin, myoglobin and cytochrome c may act as lipid oxidation catalysts in animal tissues, where they occur in high concentrations. In plants, peroxidase and catalase are significant haemin compounds. In enzyme-active plant tissues the enzyme lipoxygenase is far more active than haemoproteins as a lipid oxidation catalyst. Haemoproteins accelerate the homolytic decomposition of hydroperoxides and the iron protoporphyrins are more effective than Fe^{3+} alone.

2 Antioxidants

Antioxidants include any substance which is capable of delaying, retarding or preventing oxidative deterioration of lipids. Antioxidants can inhibit or retard the oxidation in two ways: either as chain-breaking antioxidants, so-called primary antioxidants, or as preventors of the initiation step, so-called secondary antioxidants

Primary Antioxidants

As chain-breaking antioxidants they can interfere either in the chain propagation step by scavenging lipid radicals or in the initiation step by scavenging non-lipid radicals:

$$ROO^{\bullet} + AH \rightarrow ROOH + A^{\bullet} \qquad (4)$$

$$A^{\bullet} + ROO^{\bullet} \Rightarrow \text{non-radical products} \qquad (5)$$

$$A^{\bullet} + A^{\bullet} \Rightarrow \text{non-radical products} \qquad (6)$$

$A^{\bullet}$: antioxidant radical $ROO^{\bullet}$: peroxyl radical

The free radical $A^{\bullet}$ is stabilised by either kinetic (bulky side chains) or thermodynamic (stabilised by resonance) factors and will not participate in the propagation step: an example is the **BHT** radical which is stabilised by both kinetic and thermodynamic factors. Phenolic antioxidants are usually oxidised into relatively stable aryloxyl radicals by reactions with $ROO^\bullet$ and $RO^\bullet$:

$$\text{(7)}$$

The lifetime of phenoxyl radicals varies from several seconds to several days, depending on inhibitor structure. Phenoxyl radicals having a free *ortho* or *para* position are able to dimerise.

When antioxidants react as chain-breaking antioxidants they must scavenge the hydroperoxide radical at a faster rate than they can react with another unsaturated fatty acid.[4] The reverse reaction, whereby the antioxidant radical converts the lipid peroxide to a peroxyl radical, should be slow:[5]

$$AH + LOO^\bullet \rightarrow A^\bullet + LOOH \tag{8}$$

$$A^\bullet + LOOH \rightarrow LOO^\bullet + AH \tag{9}$$

$$A^\bullet + LH \rightarrow AH + L^\bullet \tag{10}$$

AH: antioxidant molecule; LOOH: lipid peroxide molecule

The efficiency of an antioxidant depends on the ratio of the rates of reaction (8) to those of (9) and (10). A compound which is capable of reducing the antioxidant radical ($A^\bullet$) back to the parent compound (AH) will compete with reactions (9) and (10) and therefore increase the efficiency of peroxyl radical scavenging (8). The steady state concentration of the initial radical will be maintained at its initial concentration for a longer period and therefore it will result in a more efficient suppression of the peroxidation reaction. This is an example of a synergistic enhancement of antioxidant activity.

Secondary Antioxidants

Preventive inhibitors decrease the rate of autoxidation by suppressing the rate of the initiation reactions. These include the following type of inhibitors:[6]

1. Metal chelators (such as citric acid, ascorbic acid, etc.)
2. Enzyme inhibitors (such as catalase, superoxide dismutase (SOD) or gluthathione peroxidase)
3. Singlet oxygen quenchers
4. Substances preventing initiation by light or other radiation such as carotenes, which neutralise the photosensitising effect of chlorophyll

5. Substances decomposing hydroperoxides into inactive products such as various thiols and sulfides (reducing agents)
6. Substances diminishing the sensitivity of double-bond systems to free-radical formation (*e.g.* linoleic acid may be stabilised by potassium iodide by complex formation)

Natural antioxidants often fulfil a number of these functions (multiple mode of action).[7] As an example of a synthetic antioxidant, propyl gallate, a partially water-soluble antioxidant used in the food industry, is a chain-breaking antioxidant, a powerful scavenger of hydroxyl radicals and an iron chelator.

Sometimes the already mentioned synergism can be observed when two antioxidants are used together. A mixture of two antioxidants may have a much better effect than either of the components alone. If a chain-breaking and a preventive antioxidant are mixed, both initiation and propagation are suppressed. A well-know example of synergism is the system vitamin E and vitamin C, where vitamin E acts as a chain-breaking antioxidant and vitamin C reduces the formed radical back to its original state.

The critical concentration of an antioxidant is its concentration in an oxidising system necessary to interrupt all oxidation chains formed by initiation processes.[6] Satisfactory inhibition is obtained only if the concentration of antioxidant becomes higher than the critical value. At antioxidant levels higher than the critical value the concentration of free radicals in the reaction mixture soon attains a stationary value and remains constant for the induction period whereas the antioxidant is slowly used up.

A gradual reduction in the increase of antioxidant activity is observed with increasing concentration of antioxidant until a concentration is attained where further addition of antioxidant has no additional effect on the stability against autoxidation. If the concentration of antioxidant is further increased, an inversion of activity takes place and the stability decreases (pro-oxidant effect). That concentration depends on the kind of antioxidant, the type of materials stabilised and on the test conditions.

3 Biological Importance of Free Radical Reactions

It has been thought for some years that free radicals and reactive oxygen species play an important role in human disease processes including cancer, arteriosclerosis, rheumatoid arthritis, inflammatory bowel disease, immune system decline, brain dysfunction, cataracts and malaria.[8,9]

The formation of excessive concentrations of reactive oxygen species in the wrong locations can cause tissue degeneration and other harmful effects although the generation of reactive oxygen species is an essential defence mechanism of the body.[10] Reactive oxygen species of physiological significance include singlet oxygen, superoxide anion ($O_2^{\cdot}$), hydrogen peroxide and hydroxyl radicals,

Human plasma contains numerous lipid particles, the lipoproteins, which carry lipids such as phospholipids, cholesteryl esters and triacylglycerols around the body. The major lipoproteins are low-density lipoproteins (LDL), high-density lipoproteins (HDL), very-low-density lipoproteins (VLDL) and chylomicrons. Oxidation of LDL is considered a key step in the development of atherosclerosis. Oral supplementation with α-tocopherol leads to an increase in plasma and LDL tocopherol, and the oxidation resistance of the LDL from a single donor increases with α-tocopherol content or with the total LDL antioxidants. Hence, the link between dietary antioxidants and LDL stability has been established.[11] Vitamin E is the main antioxidant in LDL but other molecules also contribute to the oxidative stability of LDL including carotenoids in LDL and ascorbic acid in the aqueous phase.

4 Structure and Occurrence of the Flavonoids

Flavonoids occur in a variety of fruit, vegetables, leaves and flowers.[12] They are found as aglycones, glycosides and methylated derivatives. The most important sub-groups are the colourless catechins, the red-blue coloured anthocyanidins, the brightly yellow flavonols and flavones as well as the colourless proanthocyanidins.[13]

All the flavonoids have two distinct units (Figure 3): a C_6–C_3 fragment that contains the B-ring and a C_6 fragment (the A-ring).[14] The flavonoid aglycone normally consists of a benzene ring (A) condensed with a six-membered ring (C) which carries a phenyl ring as a substituent in the 2-position (B). The six-membered ring condensed with the benzene ring is either a γ-pyrone ring (flavonols and flavones) or its dihydro derivatives (flavanols and flavanones).

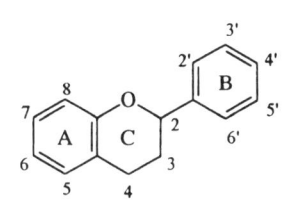

Figure 3 *Basic flavonoid structure*

The position of the benzenoid substituent on the C-ring divides the flavonoid class into flavonoids (2-position) and isoflavonoids (3-position). Flavonols differ from flavanones by a hydroxyl group in the 3-position and a C2–C3 double bond. Anthocyanidins differ from the other flavonoids in having a charged oxygen in the C-ring. The C-ring is open in the chalcones. Flavonoids are often hydroxylated in position C-3, 5, 7, 3', 4' or 5'. The structural features of the flavonoid subgroups are shown in Figure 4.

Flavonoids occur in leafy vegetables and fruit as glycosides in nature. Mono- and di-glycosides predominate but others can be found in small concentrations.[15] When glycosides are formed, the glycosidic linkage is normally located in position 3 or 7 and the carbohydrate can be L-rhamnose, D-glucose, glucorhamnose, galactose, arabinose, D-xylose or D-apinose.[16]

Antioxidant properties of flavonoids have been ascribed mainly to flavonols,

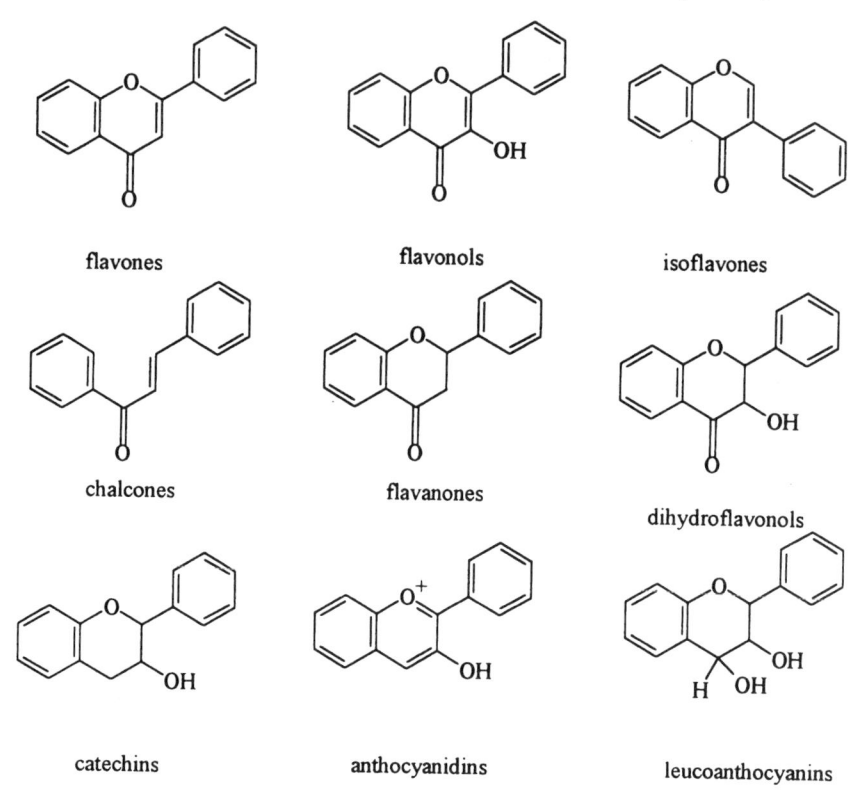

Figure 4 *Structures of flavonoid classes*

flavones and catechins. Flavonols differ from flavones in having an hydroxyl group at position 3.[17] The only difference between individual flavonols and flavones is the number of hydroxyl groups in the B-ring. They have structures as shown in Figure 5.

The preferred bonding site of the sugar molecule to the flavonol is the 3-position, much less frequently the 7-position. Diglycosides (3-*O*-biosides and 3,7-di-*O*-glycosides) can occur. Flavones occur mainly as 7-*O*-glycosides. Occasionally, C-glycosides can occur. In this case the sugar C-atom is directly attached to the aromatic ring.[18]

The formation of flavone and flavonol glycosides depends normally on the action of light,[19] so that in general the highest concentration of these compounds occur in leaves or in the skins of fruits while only traces are found in parts of plants below the ground. The common onion is, however, a well-known exception. It contains a high amount of quercetin in different layers. Mostly it is found that the higher the irradiation while growing the higher the amount of flavonoids.

Flavones and flavonols do not contribute markedly to the coloration of the plant except where they occur in high concentrations, as in the skin of onions, or when they are in a metal complex. Flavonols, which make an essential

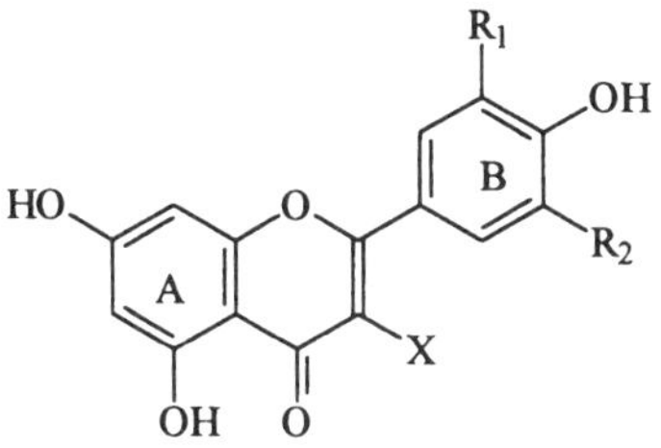

flavonol: X=OH $R_1, R_2 = H$ kaempferol
                  $R_1 = OH, R_2 = H$ quercetin
                  $R_1, R_2 = OH$ myricetin
flavone: X=H $R_1, R_2 = H$ apigenin
                  $R_1 = OH, R_2 = H$ luteolin

Figure 5 *Structures of flavonols and flavones*

contribution to the yellow colour of a flower, differ from the usual hydroxylation pattern by an additional hydroxyl group attached to the nucleus at the 6- or 8-position. Several functions of flavonoids in plants have been either demonstrated or proposed. These include: protection of plants from UV light, insects, fungi, viruses and bacteria; pollinator attractors; plant hormone controllers; and enzyme inhibitors.[20] Their significance in food has been widely investigated.[15]

Flavonols and flavones are stable against heat, oxygen, dryness, and moderate degrees of acidity, but are more or less quickly destroyed by illumination.[16]

Quercetin glycosides predominate in vegetables or in leaves of various vegetables (mostly < 10 mg/kg fresh weight) (weight always refers to the aglycone) and in fruits (average 15 mg/kg fresh weight). Frequently glycosides of kaempferol, luteolin and apigenin are also present in vegetables. In fruits, flavones have been detected rarely and in trace quantities. Kaempferol can be found in fruits in smaller quantities than quercetin. Myricetin is found in blackcurrants, black grapes, cranberries, bilberries, and broad beans,[21] and it is a significant component in red wine and grape juice. It is often present in higher concentrations than quercetin (7–9 mg/L against quercetin 4–16 mg/L).[21] Luteolin can be found (13-31 mg/kg) in vegetables like red pepper. Luteolin and apigenin glycosides occur in carrots, celery and salad.[13] Quercetin levels in the edible part of most vegetables were generally below 10 mg/kg except for onions, kale, broccoli and beans. Kaempferol could only be detected in kale, endive, leek and turnip tops.

Tea contains quercetin and kaempferol in concentrations exceeding 1% of its dry matter along with 25% catechins.[13] In black tea, myricetin could also been found but no apigenin or luteolin.[22] In green tea the major flavonoid is also quercetin (14–23 mg/L), followed by kaempferol[21] and myricetin. Apigenin and luteolin are also present.[16] In fruits, apples contain the highest amounts of quercetin. The glycoside composition of quercetin glycosides in apples is given below:

Quercetin glycosides	Weight (mg/kg)
Que-3-glucoside[23]	3–16
Que-3-galactoside[13]	5–39
Que-3-rhamnoside[24]	3–19
Que-rutinoside	1–10
Que-arabinoside	8–25
Que-xyloside	4–10

Hertog et al.[21] quoted 21–72 mg/kg as the aglycone concentration in the edible parts of the apple; kaempferol was below 2 mg/kg. The following shows the agglomeration of quercetin close to the surface in, e.g. Cox's Orange apples:[19]

Outer parts of tissue (skin, peel)		Remaining tissue	
Quercetin (mg/kg)	Kaempferol (mg/kg)	Quercetin (mg/kg)	Kaempferol(mg/kg)
263	7	<1	<0.1

As an example for a high quercetin content in vegetables the composition of onion is given. In the brown non-edible outer scales, most of the quercetin is found as the free aglycone (67–85% of total quercetin). In the epidermis, que-4'-glucoside, que-3,4'-diglucoside and que-4',7-diglycoside can be found.[25,26] Hertog[21] quoted 284–486 mg/kg quercetin in the edible parts and a kaempferol concentration below 2 mg/kg. Processing of onions reduced the flavonoid concentration to about 50% of that present in the fresh product.

5 Dietary Intake of Flavonoids

Kuehnau[16] estimated the total dietary intake of all flavonoids in the USA to be about 1 g per day (as glycosides), but this appears to be an overestimate owing to analytical inaccuracies and incomplete food composition data.[27,28] According to Kuehnau, beverages (cocoa, cola, coffee, beer and wine) contribute about 25–30%, with fruits contributing about 40% of the dietary intake of flavonoids. Vegetables, including herbs and spices, supply an extra 15%. Of the daily intake, flavones and flavonols were consumed at a level of 100 mg. Hertog et al.[29] calculated the average intake of flavonols and flavones to be about 23 mg/day (as aglycones). Quercetin is predominant with 16 mg/day. Main sources in the Netherlands are black tea (48%), onions (29%) and apples (7%).

The flavone and flavonol intake in the UK was estimated to be 30 mg/day; quercetin accounts for 64% of the total. The flavonol and flavone intake varies a lot worldwide; the highest was found in Japan (64 mg/day), with the lowest in Finland at 6 mg/day (Seven Country Study[28]). It also has to be considered that not all the different flavonoids show the same biological activity. Absorption of flavonols is mainly as glycosides.[30,31] An unknown amount of flavonoids is catabolised by intestinal bacteria.

6 Antioxidant Properties of the Flavonoids

Apart from fat-soluble tocopherols, flavonoids are the most common and most active antioxidants in foods.[16] Unlike tocopherols they can act both in hydrophilic and in lipophilic systems. Flavonoid glycosides act in hydrophilic systems but they are not as active as their aglycone to scavenge radicals.[12,32]

The flavonoids can act as primary and secondary antioxidants. The primary antioxidative activity of the flavonoids involves scavenging of superoxide anions,[33,34] hydrogen peroxide,[35] singlet oxygen[36,37] and lipid peroxide radicals[38,39] and activity against hydroxyl radicals has been reported.[32,37,40] As secondary antioxidants they can act as metal chelators[41,42] and may function as UV-light filter protectors, based on their stability in UV light under aerobic conditions and their localisation in the epidermis.[12]

Effect of Flavonoid Structure on Antioxidant Activity

There are some contradictory reports in the literature about the structural requirements of flavonoids for optimal antioxidant activity. The origin of the contradictions appear to arise from several sources. Some methods specifically measure radical scavenging activity, while other methods are dependent on flavonoid stability at high temperatures or during prolonged storage. The effect of the oxidation medium is also important, since the solubility of flavonoids in oil and water is limited, and they may often act at interfaces. The ability of flavonoids to chelate metals is affected by medium and pH. The pH also affects flavonoid stability, since flavonoids become less stable at high pH. Concentration and temperature also affect the antioxidant activity of flavonoids.

The antioxidant activity is mainly due to the phenolic hydroxyl groups, especially the 4' and 3'-hydroxyl groups in the B-ring of flavonols and flavones.[43,44] Another important factor is the presence of C-4 carbonyl and the 3-hydroxyl group.[45] The 3-hydroxyl group is a good electron donor owing to its conjugation with the B-ring but there is very little conjugation between the A- and the B-rings.[46] Glycosylation of 3-OH of quercetin to rutin has been reported to either reduce the antioxidative activity[47,48] or have no effect[49] on the antioxidant activity. It was claimed that the hydroxyl group in position 5 plays an important role,[7] but a more recent study refutes this.[44] Addition of an OH in position 8 slightly enhances the activity.[7] The *meta*-hydroxyl substitution of the A-ring is not of major importance.[50] The necessity of a 2,3-double bond, giving a fully conjugated system through the B- and C-rings, for antioxidant activity has been controversial.[7,51–53] However, recent work seems to confirm the importance of the 2,3-double bond for optimal electron donating ability.[46] The antioxidant activity has been shown in numerous studies[47,50,53,54] to increase with the number of hydroxyl groups in the B-ring. Thus, normally myricetin with an extra hydroxyl group in the B-ring is more active as an antioxidant than quercetin. However, effects of medium, etc., may cause a reversal of this order in some studies. Quercetin was found to be more active than myricetin in scavenging superoxide radicals.[33] The most important functions for radical scavenging[55] are shown in the case of quercetin in Figure 6.

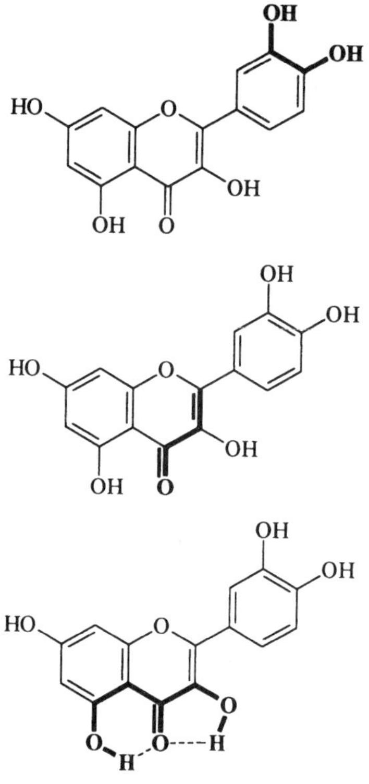

Figure 6 *Functional groups associated with radical scavenging activity of quercetin*[55]

Bors et al.[55] studied the rate at which flavonoids capture radicals such as hydroxyl, azide ($N_3^\cdot$), superoxide-, linoleic acid peroxyl (LOO$^\cdot$), *tert*-butoxyl (*t*-BuO$^\cdot$)- and sulfite (SO$_3^\cdot$). This study showed that most flavonoids have high rate constants in these reactions. Both kaempferol and quercetin are highly efficient radical scavengers; however only the quercetin aroxyl radical decays slowly enough to make it a potent antioxidant. Bors concluded that it was necessary to have an *o*-dihydroxy (catechol) structure in the B-ring, a 2,3-double bond in conjugation with 4-keto group and additional 3- and 5-OH groups for optimal radical-scavenging potential. At pH 11.5 a doubly dissociated state of quercetin can be expected involving the dissociation of 7- and 4'-OH (dissociation sequence is 7-OH before 4'-OH before 5-OH) (Figure 7).[61]

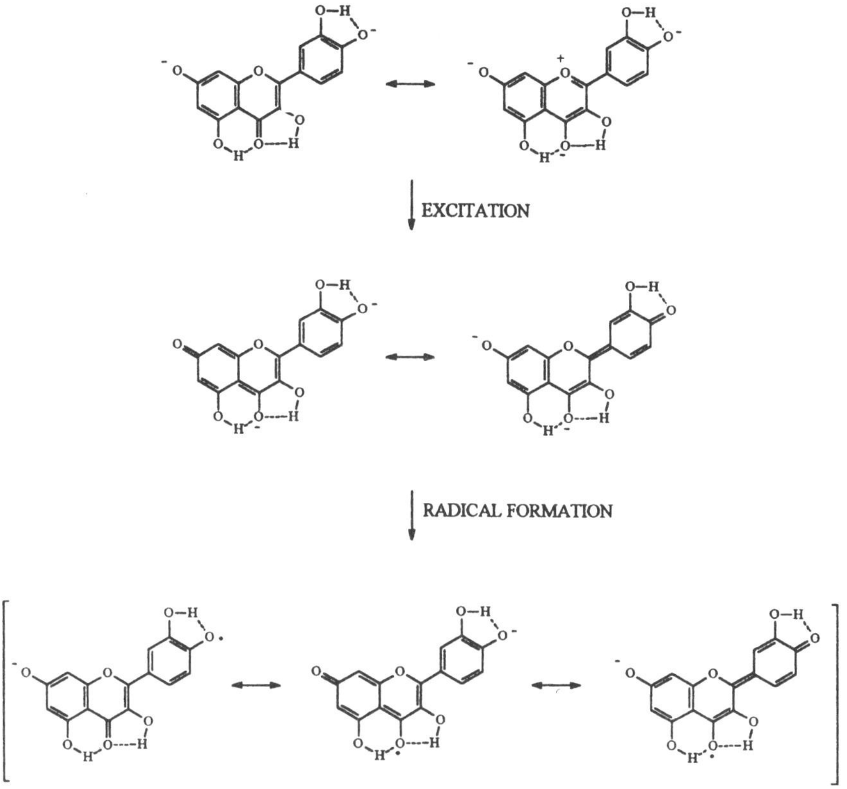

Figure 7 *Formation of radicals at high pH (11.5)*[56]

During the scavenging of active species of oxygen, the flavonols are oxidised.[32] A mechanism for the oxidation of kaempferol has been suggested.[12] Kaempferol can be oxidised to 4-hydroxyphenylglyoxylic acid and phloroglucinolcarboxylic acid in the presence of superoxide anion at pH 8 (Figure 8)[57].

Certain flavonoids show synergistic effects of which the synergy between vitamin C and flavonoids is well established. It is particularly strong with catechins, flavonols and flavones.[58] This synergy has three possible mechanisms: the first is simple oxidation-inhibition effects; the second involves the chelation of copper by the flavonoids; the third mechanism suggests a charge-transfer complex between vitamin C and flavonoids.

Metal Chelation by Flavonols

Metals may chelate with the 4-carbonyl and 5-hydroxyl groups or with *ortho*-dihydroxyl groups in the B-ring. Aluminium chloride is commonly used as a shift reagent for identifying the structure of flavonols, since a large bathochromic shift occurs with both these pairs of substituents, but the addition of

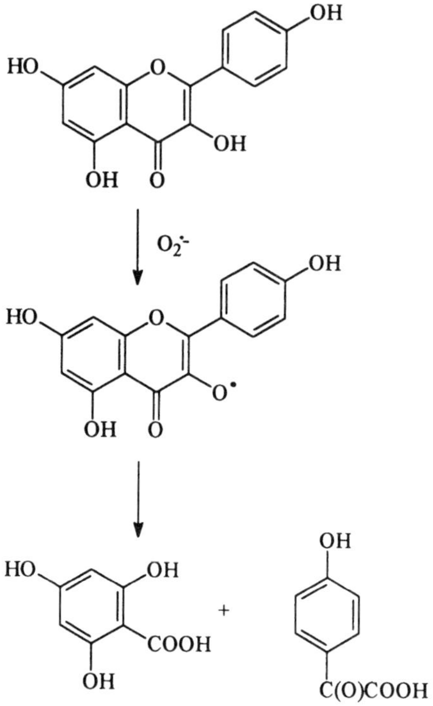

Phloroglucinolcarboxylic acid 4-Hydroxyphenylglyoxylic acid

Figure 8 *Oxidation of kaempferol by superoxide anion at pH 8*[57]

strong acid only removes the chelation with dihydroxyl substituents.[59] Addition of 5 μg copper to 2 g flavonols from onions (mainly quercetin) resulted in the chelation of 56% of the copper.[60] Chelation of quercetin with iron appears to be stronger than with copper, with the free quercetin band at 373.7 nm almost completely removed by the addition of an equimolar concentration of a ferric salt, whereas quercetin and copper form a labile complex.[44]

Effect of Medium on Antioxidant Activity

The medium where the oxidation takes place also plays an important role in determining the antioxidant activity of flavonoids. Chen *et al.* [61] claimed that myricetin was the most active antioxidant of several flavonoids in canola oil, whereas quercetin was the most active flavonoid in an emulsion and quenched free radical chain reactions most efficiently in rat blood cell membranes. However, the use of metal ions in the emulsion studies but not in the oil experiments described in this paper makes it uncertain whether the differences were simply a medium effect.

Quercetin was more active as an antioxidant than myricetin when assessed by the scavenging of the 2,2′-azinobis(3-ethylbenzothiazoline-6-sulfonic acid)

(ABTS) radical cation in aqueous solution,[62] whereas a reverse order of activitity has been found in oils[63] and emulsions.[64] Barrera-Arrelano and Esteve[65] also observed differences in the activity of apigenin and rutin in aqueous systems versus lipid systems. Quercetin was highly active in a micellar system[66] and in liposomes.[67] Quercetin was a much more effective antioxidant than its glycoside rutin, when assessed by β-carotene oxidation in an emulsion, but the difference was much smaller when assessed by scavenging of the α,α-diphenyl-β-picrylhydrazyl (DPPH) radical scavenging method in methanol.[50]

Pro-oxidant Activity of Flavonoids

Pro-oxidant effects of flavonols have been observed when flavonols have been studied at alkaline pH in the presence of metal ions.[68–71] These effects have been ascribed to the production of hydrogen peroxide, superoxide anion and hydroxyl radicals from the flavonol.[69] Quercetin and myricetin greatly accelerated generation of hydroxyl radicals from hydrogen peroxide in the presence of ferric-EDTA chelates at pH 7.4 as measured by the deoxyribose assay.[68] Myricetin produced a semiquinone type radical above pH 8.[69,70] Semiquinones are known to reduce ferric ions to ferrous ions that can then reduce hydrogen peroxide to hydroxyl radicals. The addition of iron resulted in the disappearance of the semiquinone ESR signal and the appearance of a signal due to a radical derived from a hydroxyl radical at pH 7.5 and above.[69] In the absence of iron, radicals formed from quercetin only appeared above pH 9.

Antioxidant Activity of Quercetin Relevant to the Prevention of Coronary Heart Disease (CHD)

Robbins[72] found that citrus flavonoids may reduce aggregation of blood platelets and hence may have an effect on coronary thrombosis. Epidemiological studies indicate that there is a low incidence of CHD in some French cities despite a high intake of dairy fat, which was termed the 'French paradox'. This has been explained by the consumption of wine in the French diet,[73] and may illustrate an effect of flavonoids in the wine[74], although alternative explanations including the effect of ethanol, vitamin E or lifestyle have also been proposed.[75]

The Zutphen Elderly Study[76] was another epidemiological study that indicated that the dietary intake of flavonoids by elderly men correlated with reduced CHD mortality. Relative risks of mortality from CHD were about 50% lower in the highest tertile of flavonoid intake (>29.9 mg/day) than in the lowest tertile (<19 mg/day) after adjustment for dietary variables and non-dietary risk factors. However, a subsequent study on a Welsh population failed to find a similar effect.[77]

Quercetin has been shown to be a potent inhibitor of the *in vitro* oxidation of LDL both by macrophages and copper sulfate.[78] Tocopherols within the LDL are also conserved by the presence of quercetin. However, myricetin is unlikely to be beneficial by its action in LDL, since it is capable of the nonoxidative modification of LDL.[79] Vinson et al.[80] found that quercetin was more active than myricetin in retarding copper-catalysed oxidation of LDL. However, flavonols have not been found in LDL, and they may act by an effect on the eicosanoid balance, since they have been shown to inhibit eicosanoid synthesis, a key step in platelet aggregation, by their action on cyclooxygenase or lipoxygenase.[81,82]

References

1. W. Grosch, *Food Flavours. Part A, Introduction*, Edited by I.D. Morton and A.J. Macleod, Elsevier, Barking, 1982, 325.
2. P.J. O'Brien, *Can. J. Biochem.*, 1969, **47**, 485.
3. F.W. Heaton, and N. Uri, *J.Lipid Res.*, 1961, **2**, 152.
4. N. Niki, M. Noguchi, and M. Iwatsuki, *Nutritition, Lipids, Health and Disease*, Edited by A.S.H. Ong, E. Niki and L. Packer, AOCS Press, Champaign, 1995, 1.
5. V.M. Darley-Usmar, A. Hersey, N. Gotch, E. Niki, N. Hogg, B. Kalyanaramn, D. Stone, and M.T. Wilson, *Antioxidants, Free Radicals and Polyunsaturated Fatty Acids in Biology and Medicine*, Edited by A.T. Diplock, J.M.C. Gutterridge and V.K.S. Skuhla, International Food Science Centre A/S, Lysterup, 1993, 49.
6. J. Pokorny, *Autoxidation of Unsaturated Lipids*, Edited by H.W.S.Chan, Academic Press, London, 1987, 141.
7. M. Namiki, *Crit. Rev. Food Sci. Nutr.*, 1990, **29**, 273.
8. M.H. Gordon, *Natural Product Reports*, 1996, **13**, 265.
9. B. Halliwell, J.M.C. Gutteridge, and C.E. Cross, *J. Lab. Clin. Med.*, 1992, **119** (6), 598.
10. B. Halliwell, M.A. Murcia, S. Chirico, and O. Aruma, *CRC Crit. Rev. Food Sci. Nutr.*, 1995, **35**, 7.
11. I. Jialal, C.J. Fuller, and B.A. Huet, *Arterioscler. Thromb. Vasc. Biol.*, 1995, **15**, 190.
12. R.J. Hsieh and J.E. Kinsella, *Adv. Food Nutr. Res.*, 1989, **33**, 233.
13. K. Herrmann, *Gordian*, 1993, **93** (7–8), 108.
14. T.A. Geissmann, *The Chemistry of Flavonoids Compounds*, Edited by T.A. Geissmann, Pergamon Press, Oxford, 1962, 1.
15. J. Gripenberg, *The Chemistry of Flavonoids Compounds*, Edited by T.A. Geissmann, Pergamon Press, Oxford, 1962, 406.
16. J. Kuehnau, *Wld. Rev. Nutr. Diet*, 1976, **24**, 117.
17. K. Herrman, *Z. Lebensm. Unters. Forsch.*, 1970, **144**, 191.
18. K. Herrmann, *Lebensmittelchem. Gerichtl. Chem.*, 1979, **33**, 4.
19. K. Herrmann, *J. Food Technol.*, 1976, **11**, 433.
20. K.P. Markham, *Methods Plant Biochem.*, 1989, **1**, 197.
21. M.G.L. Hertog, P.C. Hollmann, and M.B. Katan, *J. Agric. Food Chem.*, 1992, **40**, 2379.
22. M.G.L. Hertog, P.C.H. Hollmann, and H. Van de Putte, *J. Agric. Food Chem.*, 1993, **41**, 1242.
23. H.W. Siegelmann, *Z. Lebensm. Unters. Forsch.*, 1954, **98**, 647.
24. J. Van Buren, *The Biochemistry of Fruits and their Products*, Edited by A.C. Hulme, Academic Press, London, 1970, 269.
25. B.J. Brandwein, *J. Food Sci.*, 1965, **30**, 680.
26. H. Starke and K. Herrmann, *Z. Lebensm. Unters. Forsch.*, 1976, **161**, 137.

27. W.S. Pierpoint, *Plant Flavonoids in Biology and Medicine: Biochemical, Pharmalogical and Strucure-Activity Relationships*, Liss, New York, 1986, 125.
28. M.G.L. Hertog and P.C.H. Hollman, *Eur. J. Clin. Nutr.*, 1996, **50**, 63.
29. M.G.L. Hertog, P.C.H. Hollman, M.B. Katan, and D. Kromhout, *Nutr. Cancer,* 1993, **20**, 21.
30. G. Paganga and C.A.Rice-Evans, *FEBS Lett.*, 1997, **401**, 78.
31. P.C.H. Hollman, J.M.P. van Trijp, M.N.C.P. Buysman, M.S.v.d. Gaag, M.J.B. Mengelers, J.H.M. de Vries, and M.B. Katan, *FEBS Lett.*, 1997, **418**, 152.
32. U. Takahama, *Plant Physiol.*, 1983, **71**, 852.
33. J. Robak and R.J. Gryglewski, *Biochem. Pharmacol.*, 1988, **37**, 837.
34. U. Takahama, *Photochem. Photobiol.*, 1985, **42**, 89.
35. U. Takahama, R.J. Youngman and E.F. Elstner, *Phytobiochem. Phytobiophys.,* 1984, **7**, 175.
36. U. Takahama, *Cell. Physiol.*, 1984, **25**, 1181.
37. U. Takahama, *Plant Physiol.*, 1984, **74**, 852.
38. U. Takahama, *Photochem. Photobiol.*, 1983, **38**, 363.
39. U. Takahama, *Phytochemistry,* 1985, **24**, 1443.
40. S.R. Husain, J. Cillard, and P. Cillard, *Phytochemistry*, 1987, **26**, 2489.
41. G. Paganga, H. Al-Hasim, H. Khodr, B.C. Scott, O.I. Aruoma, R.C. Hider, B. Halliwell, and C.A. Rice-Evans, *Redox Report,* 1996, **2**, 359.
42. J.E. Brown, H. Khodr, R.C. Hider, and C.A. Rice-Evans, *Biochem. J.*, 1998, **330**, 1173.
43. F. Shahidi, Z. Yang, and O. Saleemi, *J. Food Lipids,* 1993, **1**, 69.
44. G. Cao, E. Sofic and R.L. Prior, *Free Rad. Biol. Med.*, 1997, **22**, 749.
45. D.L. Crawford, R.O. Sinnhuber, and H. Aft, *J. Food Sci.,* 1961, **26**, 139.
46. S.V. Jovanovic, S. Steenken, Y. Hara, and M.G. Simic, *J. Chem. Soc. Perkin Trans.*, **2**, 1996, 2497.
47. A. Letan, *J. Food Sci.,* 1966, **31**, 518.
48. A. von Gadow, E. Joubert and C.F. Hansmann, *J. Agric. Food Chem.*, 1997, **45**, 632.
49. B.J.F. Hudson and J.I. Lewis, *Food Chem.*, 1983, **10**, 47.
50. A.C. Mehta and T.R. Seshadri, *J. Sci. Ind. Res.,* 1959, **18B**, 24.
51. S.Z.Dziedzic and B.J.F.Hudson, *Food Chem.*, 1984, **14**, 45.
52. K. Herrmann, *Fette, Seifen, Anstrichm.,* 1973, **75**, 499.
53. A. Letan, *J. Food Sci.,* 1966, **31**, 395.
54. I. Thuman and K. Herrmann, *Dtsch. Lebensm.-Runds.*, 1980, **76**, 344.
55. W. Bors, W. Heller, C. Michel and M. Saran, *Methods Enzymol.,* 1990, **186**, 343.
56. T.J. Mabry, K.R. Markham, and M.B. Thomas, *The Systematic Identification of Flavonoids, Part 2*, Springer Verlag, Berlin, 1970.
57. U. Takahama, *Plant Cell Physiol,* 1987, **28**, 953.
58. K.AHarper, *J. Food Technol.,* 1969, **4**, 405.
59. K.P. Markham, *Techniques of Flavonoid Identification,* Academic Press, London, 1982, 36.
60. R. Eller and G. Weber, *Z. Anal. Chem.,* 1987, **328**, 492.
61. Z.Y. Chen, P.T. Chan, K.Y. Ho, K.P. Fund, and J. Wang, *Chem. Phys.Lipids,* 1996, **79**, 157.
62. C.A. Rice-Evans, N.J. Miller, P.G. Bolwell, P.M. Bramley, and J.B. Pridham, *Free Rad. Res.,* 1995, **22**, 375.
63. U.N. Wanasundara and F. Shahidi, *Food Chem.,* 1994, **50**, 393.

64. D.E. Pratt, *Phenolic, Sulfur and Nitrogen Compounds in Food Flavour,* ACS Symposium Series 26, Washington, 1976, 1.
65. D. Barrera-Arellano and W. Esteve, *Cienc. Tecnol. Aliment.*, 1989, **9**, 107.
66. M. Foti, M. Piattelli, M.T. Baratto and G. Ruberto, *J. Agric. Food Chem.*, 1996, **44**, 497.
67. J. Terao, *Proceedings of the 21st World Congress of the International Society for Fat Research,* 1996, 327.
68. M.J. Laughton, B. Halliwell, P.J. Evans and J.R.S. Hoult, *Biochem. Pharmacol.*, 1989, **38**, 2859.
69. W.F. Hodnick, F.S. Kurg, W.J. Foettiger, C.W. Bohnmont and R.S. Paridni, *Biochem. Pharmacol.*, 1986, **35**, 2345.
70. A.T. Canada, E. Giranella, T.D. Ngugen and R.P. Mason, *Free Rad. Biol. Med.*, 1990, **9**, 441.
71. S.C. Sahu and G.C. Gray, *Cancer Lett.*, 1996, **104**, 193.
72. R.C. Robbins, *Citrus Nutrition and Quality*, ed. S. Nagy and J.A. Attaway, ACS Symposium Series no.143, American Chemical Society, Washington DC, 1980, 23.
73. S. Renaud and M. de Lorgeril, *Lancet*, 1992, **339**, 1523.
74. E.N. Frankel, J. Kanner, J.B. German, E. Park and J.E. Kinsella, *Lancet,* 1993, **341**, 454.
75. M.C. Bellizzi, M.F. Franklin, G.G. Duthie and W.P.T. James, *Europ. J. Clin. Nutr.*, 1994, **48**, 822.
76. M.G.L. Hertog, E.J.M. Fresken, P.C.H. Hollman, M.B. Katan and D. Kromhout, *Lancet*, 1993, **342**, 1007.
77. M.G.L. Hertog, P.M. Sweetnam, A.M. Fehily, P.C. Elwood and D. Kromhout, *Am. J. Clin. Nutr.*, 1997, **65**, 1489.
78. C.V. De Whalley, S.M. Rankin, J.R.S. Hoult, W. Jessup and D.S. Leake, *Biochem. Pharmacol.*, 1990, **39**, 1743.
79. S.M. Rankin, C.V. De Whalley, R.S. Hoult, W. Jessup, G.M. Wilkins, J. Collard and D.S. Leake, *Biochem. Pharmacol.*, 1993, **45**, 67.
80. J.A. Vinson, Y.A. Dabbagh, M.M. Serry and J. Jang, *J. Agric. Food Chem.*, 1995, **43**, 2800.
81. C.R. Pace-Asciak, S. Hahn, E.P. Diamandis, G. Soleas, and D.M. Goldberg, *Clin. Chim. Acta*, 1995, **235**, 207.
82. R. Landolfi, R.L. Mower, and M. Steiner, *Biochem. Pharmacol.*, 1984, **33**, 1525.

5
Dietary Fatty Acids, Postprandial Lipaemia and Coronary Heart Disease

Christine M. Williams

HUGH SINCLAIR UNIT OF HUMAN NUTRITION, DEPARTMENT OF FOOD SCIENCE AND TECHNOLOGY, UNIVERSITY OF READING, PO BOX 226, WHITEKNIGHTS, READING RG6 6AP, UK

1 Introduction

Lipids circulate in the plasma as lipoprotein particles of varying density, according to the nature and amount of lipid and protein present in the particle (Table 1). Cholesterol is carried in two main types of particle, low density lipoprotein (LDL) and high density lipoprotein (HDL), with high levels of the former linked to increased, and of the latter linked to decreased, risk of coronary heart disease (CHD). Triglycerides are transported in chylomicrons (CMs), very low density lipoproteins (VLDL) and also in intermediate density lipoprotein (IDL), which is a circulating product of CM and VLDL catabolism. These are collectively known as the triglyceride rich lipoproteins (TRL). Until recently, little attention has been paid to potential effects of diet on CMs and VLDL because of lack of evidence from epidemiological studies linking elevated fasting levels of this lipid fraction to risk of CHD. The strong epidemiological and biochemical evidence linking elevated levels of LDL cholesterol to risk of CHD has meant that dietary recommendations for fatty acid intakes aimed at reducing the prevalence of CHD are largely based on the cholesterol hypothesis and on the well established effects of specific fatty acid classes on fasting total cholesterol, LDL and HDL concentrations.[1,2] However recent research has suggested involvement of triglyceride-rich lipoproteins (TRL) in the pathogenesis of both atherosclerosis and thrombosis[3-7] and indicates a need to understand the effects of dietary fatty acids on these lipoproteins, concentrations of which are markedly altered following meal fat ingestion. This research has also revealed the paucity of dietary studies which evaluate postprandial as well as fasting lipoproteins and the need for more information in this area.

Table 1 Major lipoprotein fractions in human plasma

Fraction	Major lipid
Chylomicrons	Triglyceride
VLDL	Triglyceride
IDL	Triglyceride/ cholesterol
LDL	Cholesterol
HDL	Cholesterol

2 Postprandial Triglyceride Metabolism—an Outline

The postprandial TRL are the CMs, which carry triglycerides of exogenous, dietary origin, and VLDL, which carry triglycerides formed *de novo* in the liver. Following consumption of a fat-containing meal, triglyceride concentrations increase from fasting levels of approximately 1.0–1.5 mmol/l, rising to postprandial concentrations in the region of 1.5–4.0 mmol/l. Levels reach peak values between 3 and 6 hours after a meal, with the extent of the rise depending upon both the amount and type of fat in the meal. Because most people eat sequentially throughout the day, triglyceride concentrations are raised above fasting values for 16–20 hours of each 24 hour period, which means that the postprandial state is the normal nutritional state as far as lipoprotein metabolism is concerned. Cohn and co-workers have estimated that over 80% of the postprandial increase in triglyceride is due to the influx of intestinally-derived chylomicrons.[8] A small increase in large VLDL is seen in the first few hours after the meal, but this is not sustained and levels fall below baseline within six hours of meal consumption. Essentially similar conclusions were drawn by Schneeman *et al.*[9] who used a specific antibody to separate chylomicrons and VLDL.

Triglycerides in CMs and VLDL are rapidly removed from the circulation through the action of lipoprotein lipase (LPL), present on adipose tissue, mammary gland and skeletal and cardiac muscles. Fatty acids released following hydrolysis of CM and VLDL triglyceride are rapidly taken up and re-esterified in adipose tissue, whereas in skeletal and cardiac muscle the fatty acids provide an important source of energy. The development of a technique for studying substrate exchange across abdominal adipose tissue *in vivo*, pioneered by the work of Frayn and colleagues, has made an important impact on the understanding of lipid disposition in fasting and postprandial states. Frayn[10] measured arteriovenous differences across adipose tissue of CM and VLDL triglyceride following meal consumption. A marked increase in chylomicron triglyceride extraction was seen within 60 minutes of meal consumption and was associated with a corresponding reduction in VLDL triacylglycerol removal. These data support the view that these particles are competing for removal by the regulatory enzyme, lipoprotein lipase, and that this competition may be an important determinant of postprandial TRL response to meal consumption.

Thus the factors which influence postprandial triglyceride concentration will be the rate of entry *via* CMs (diet) and VLDL (liver), and the rate of removal by LPL. Dietary fatty acids have the potential to influence all three processes through their effects on CM and VLDL synthesis and, on the basis of recent work conducted in our own and other laboratories, on the activity and gene expression of LPL.

3 Postprandial Triglyceride-rich Lipoproteins and the Atherogenic Lipoprotein Phenotype

Recent research indicates that subjects with CHD[6,11] and the offspring of parents with CHD[12] show an exaggerated response to a fat-containing meal. Associated with this phenomenon is an increased prevalence of low HDL and a greater proportion of LDL in the small, dense atherogenic form. This abnormality has been linked to insulin resistance but may occur in the absence of frank diabetes and appears to be a consequence of a primary defect in triglyceride, rather than cholesterol, metabolism. Notably these lipid characteristics are also frequently observed in type 2 diabetics, who although at greater risk of CHD, frequently show total and LDL cholesterol levels in the normal range. Various names have been given to this lipid disturbance including the 'atherogenic lipoprotein phenotype (ALP)' and, in its more extreme insulin-resistant form, 'metabolic syndrome' and 'syndrome X'. It is estimated that the ALP may occur in 20–25% of the normal middle aged male population, and because the particular lipid abnormalities involved are infrequently measured as part of routine lipid screening, may go undetected in a significant proportion of those affected.

Although the detailed mechanisms which link an exaggerated lipaemic response, low HDL and raised small dense LDL with defects in insulin sensitivity are beyond the scope of this review, a brief outline of the relationships between CMs and VLDL, and the cholesterol-rich lipoproteins, is given here to illustrate one view of how an elevated postprandial lipaemic response may be related to other lipid abnormalities of the 'ALP'. The processes thought to be involved in the relationship between triglyceride-rich and cholesterol-rich lipoproteins, and which may explain the link between elevated triglycerides and accelerated atherogenesis, have been reviewed elsewhere[13,14] with an outline summary of these events shown in Figure 1.

An exaggerated response to a fat-containing meal may be due to overproduction of CMs or VLDL, or may reflect slower clearance of these particles by the rate regulatory step in triglyceride hydrolysis and removal, which is mediated by lipoprotein lipase (LPL). Accumulation of partially-hydrolysed remnant particles may also contribute to the elevated triglyceride response, since owing to their poor recognition by hepatic receptors there is reduced uptake of these cholesterol-enriched particles by receptor mediated pathways. The net consequence of these defects is increased retention of TRLs within the circulation, providing greater opportunity for neutral lipid exchange, catalysed by cholesterol ester transfer protein (CETP), which mediates the reciprocal transfer of triglyceride and cholesterol between TRLs and HDL and LDL. The net consequence of these transfers is triglyceride accumulation on HDL and LDL, and cholesterol on CM and VLDL and their remnants. The triglyceride-

1. NORMAL LIPAEMIA

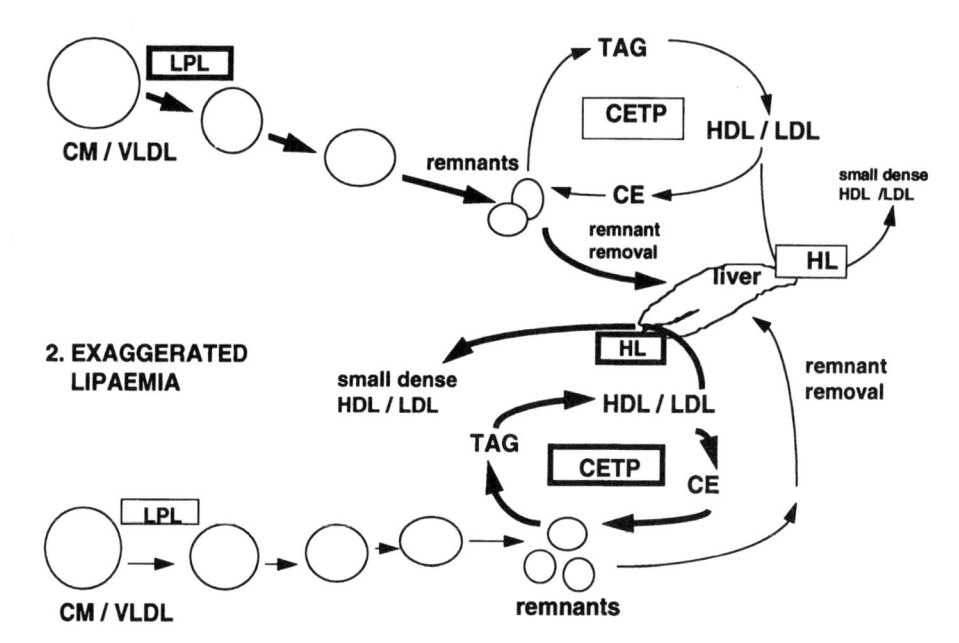

Figure 1 *Diagrammatic representation of the atherogenic consequences of impaired triglyceride tolerance. 1. Normal lipaemia : TRL production rates and clearance and remnant removal rates shown as bold lines and exchange with HDL and LDL and HL action shown as faint lines. 2. Exaggerated lipaemia: clearance of TRL and removal of remnants shown as faint lines and exchange with HDL and LDL and HL action shown as bold lines*

enriched HDL and LDL act as good substrates for the hepatic lipase (HL) enzyme, resulting in formation of small dense HDL (HDL3) and LDL (LDL3). The former is catabolised rapidly resulting in reduced circulating HDL levels. The small dense LDL and the cholesterol-enriched remnant particles are reputed to have greater atherogenic potential due to their prolonged retention within the circulation[15,16] their ability to induce foam cell formation,[17–19] and in the case of DL, to its greater susceptibility to undergo oxidation.[20] Chylomicron remnants have also been shown to penetrate the endothelium in much the same way as LDL and HDL.[21] The major lipid abnormalities of the ALP, thought to arise as a consequence of these events, are shown in Table 2

Table 2 *Lipid abnormalities of the 'atherogenic lipoprotein phenotye'*

Low HDL (< 1.0 mmol/l)
Moderately raised fasting triglycerides (1.5–3.0 mmol/l)
Increased proportion of LDL as small dense form (> 45% as LDL3)
Raised concentrations of remnant particles
Elevated postprandial triglyceride response to standard meal

As described above and outlined in Figure 1, it appears to be the elevated levels of TRL which drive the events leading to accumulation of atherogenic lipoproteins (remnants, small dense LDL, low HDL). It is therefore important to consider the role of dietary fatty acids in modulating postprandial triglyceride response since, at present, dietary recommendations do not take into consideration the effects of diet on this lipid fraction, despite evidence that disturbances in TRL metabolism are common in middle aged subjects and may increase risk of CHD.

4 Effects of Dietary Fatty Acids on Postprandial Lipaemia

In order to evaluate the effect of meal fatty acid composition on the extent of postprandial lipaemia, a number of investigators have compared triglyceride responses to single meals of varying fatty acid composition. Variations in postprandial lipaemia in response to meals of different fatty acid composition could reflect differences in the rate of synthesis, secretion or clearance of the intestinally derived CMs and the rate of production and clearance of their remnants. It is also possible that hepatic uptake of fatty acids released from chylomicrons, and the nature of the fatty acids in chylomicron remnants presented to the liver, could influence, acutely, the synthesis and secretion of VLDL, and thereby the hepatic contribution to the triglyceride response to meal consumption. Although some studies have attempted to address the relative contributions of CM and VLDL, the limitation of methods for measuring CMs or VLDL separately[22,23] make it difficult to draw conclusions regarding the involvement of CM and VLDL in fatty acid-induced variations in postprandial lipaemia. Similar limitations apply to studies which have attempted to address the question of adaptational responses to diets of different fatty acid composition. In these studies, subjects fed diets for periods ranging from 2 to 8 weeks are usually given test meals of standard fatty acid composition to evaluate effects of background diet on postprandial lipaemia. In these models, any effect of diet could reflect adaptational responses in enterocytic and hepatic enzymes involved in CM and VLDL synthesis, as well as in the lipoprotein and hepatic lipases which regulate clearance of triglyceride and remnant particles respectively. Other possible loci for adaptational effects of dietary fatty acids include receptors, such as the remnant and LDL receptors which regulate the uptake and removal of CM and VLDL remnant particles.

Although a number of animal feeding studies have used tracer techniques and measurements of tissue enzymes and receptor activities to investigate mechanisms involved in effects of specific fatty acids on CM metabolism,[24,25] it is likely that most useful information will be gained from studies investigating human TRL responses to dietary fatty acid modification.

Meal Fatty Acid Composition and Postprandial Lipaemia

There are five studies reported in the literature which have measured postprandial lipaemic responses to meals of different fatty acid composition[26-30] and one which has studied combined effects of both meal and background diet.[31] Comparison of the findings from these studies is difficult, because even when similar comparisons are being made, e.g. SFA (saturated fatty acids) vs. $n-6$ PUFA (polyunsaturated fatty acids), the nature and amounts of the fats and oils used have not always been the same and the duration of postprandial follow-up after the meal is frequently different.

In 1988, Weintraub and co-workers compared postprandial lipoprotein responses to meals comprising predominantly saturated (SFA) or $n-6$ PUFA. No differences in postprandial retinyl palmitate or triglyceride responses to the two meals were observed.[28] This is in slight contrast to our findings,[29] in which subjects were given three test meals which were identical apart from the oils used which were mixed (SFA), corn ($n-6$ PUFA) and fish oil (long-chain $n-3$ PUFA). The mixed oil was formulated to represent the current UK dietary fatty acid intake. Postprandial triglyceride response to the $n-6$ PUFA (corn oil) meal was lower than to the mixed oil meal but the major difference found was in response to the meal containing long-chain $n-3$ PUFA (fish oil), where marked attenuation in postprandial triglyceride levels was seen. In an earlier study (1988), Harris et al. found no difference in postprandial triglyceride response to liquid test meals containing either a mixture of peanut oil and cocoa butter (saturated fat meal) or Max EPA (fish oil meal).[27] A recent study[32] has shown that addition of only small amounts of long-chain (LC) $n-3$ PUFA to a standard fat-containing test meal can significantly reduce the postprandial lipaemic response compared with the same meal without added LC $n-3$ PUFA, and suggests that the fatty acids in fish oil alter the rate of synthesis and/or removal of chylomicron triglyceride, independently of the remaining fatty acid substrate supplied.

There is also some disparity in the findings of studies which have compared the effects of monounsaturated fatty acids and other major fatty acid classes on postprandial lipaemia. Bruin and co-workers[26] reported that lipaemic response was slightly higher when lipid emulsions containing olive oil were fed than when cream or sunflower oil emulsions were fed. However, we found no significant difference in plasma or chylomicron fraction triglyceride, apo B-48 or retinyl ester responses when meals of varying MUFA content were fed.[33,34] The three test meals fed contained 12%, 17% and 24% respectively of MUFA, representing current UK, current Mediterranean and previous Mediterranean MUFA intakes. On the basis of this small number of studies it is concluded that similar postprandial lipaemic responses are observed when meals consisting of predominantly SFA, MUFA or $n-6$ PUFA are fed, but when meals containing significant amounts of $n-3$ PUFA are given, postprandial lipaemia is attenuated.

Effects of Background Dietary Fatty Acid Intake on Postprandial Lipaemia

The nature of the fat fed in the habitual diet appears to influence the lipaemic response to meals. Four studies used a whole diet approach in which total dietary fat was altered,[27,35,28,31] whereas another four used capsule supplements[36–39] and one used microencapsulated oils incorporated into conventional foods,[40] to partially alter background dietary fatty acid intake. One other study has compared effects of different PUFA- and MUFA-containing diets on fasting and postprandial triglyceride values but the limited data presented make the findings difficult to compare with other studies.[41]

Demacker and co-workers[31] altered the diets of free living subjects so that either butter fat or safflower oil were the predominant fats used in cooking and in meals prepared at their work place. After three weeks on each diet, subjects' responses to a breakfast, lunch and dinner were monitored over a 24 hour period. On each occasion the test meal given was the same as that of the background diet. In this type of study it is difficult to distinguish the effect of altering the background diet from that of altering the test meal fatty acid composition. However, the findings suggests that when a diet and a test meal consisting largely of $n-6$ PUFA is fed, postprandial triglyceride and apo B-48 responses are lower than when a diet and a meal rich in SFA are fed.

The study of Weintraub and colleagues[28] was carried out in a metabolic unit with subjects randomised to receive three different background diets for three weeks over a 12 week period. Subjects' responses to two different test meals were evaluated at the end of each dietary period. Subjects' lipaemic responses were reduced when the background diet consisted of $n-6$ or $n-3$ PUFA, although it was only the reduction with $n-3$ PUFA which was statistically significant when a saturated fat meal was used as the test meal. The attenuating effects of the $n-6$ PUFA diet were greater when an $n-6$ PUFA test meal was given than when an SFA test meal was given, supporting the findings[31] which indicate there may be synergism between the adaptational response and acute meal mechanism. Harris and co-workers[27] also found hypolipaemic effects of a diet rich in fish oil in a study in which they fed either a control diet consisting of peanut oil and cocoa butter or a diet consisting of salmon oil or Max EPA as background.

These studies suggest that diets enriched with LC $n-3$ PUFAs significantly improve triglyceride tolerance. However, the level at which these fatty acids need to be given is an important consideration in relation to possible future population recommendations, which, it may be argued, should be based on triglyceride-lowering as well as cholesterol-lowering properties of dietary fatty acids.

From studies conducted using capsules containing fish oil, it would appear that diets containing 1.6–2.8 g LC $n-3$ PUFA per day result in a 30% reduction in postprandial lipaemia. Higher levels than this, *e.g.* 5 g per day can achieve an even greater degree of reduction, but this level of intake is considerably higher than habitual levels found in most human diets. In the UK average levels of LC $n-3$ PUFA intake are approximately 0.2 g per day.[42] This compares with intake levels of 12 g per day for linoleic acid, the major $n-6$ PUFA, and 1.3 g per day of linolenic acid, the precursor $n-3$ PUFA.

Effects of Habitual Fatty Acid Intake on Postprandial Lipaemia

Although the dietary studies described above are designed to address the question of adaptational responses to altered diet, it is likely that these responses develop over many weeks or even months, depending on the site and mechanism involved. It is possible that many of the studies conducted to date have been too short to reveal the full extent of changes in lipoprotein metabolism which occur when fatty acid intake is markedly altered in human diets. This is particularly so in the light of the large number of sites through which fatty acids may mediate effects on triglyceride metabolism. This means that differences may not become evident in short term dietary studies where the different diets have only been fed for a limited time (usually weeks). This conclusion is supported by a recent study we have carried out which measured postprandial responses to standard meals in subjects living in Southern and Northern Europe whose diets closely reflected the typical diets of their country.[43,44] We found the pattern of lipaemic response to identical meals was very different in Southern than Northern Europeans and that, in particular, subjects from Southern Europe showed a marked early rise in triglyceride with a rapid return to fasting values, whereas Northern Europeans showed a slow sluggish rise in triglycerides which did not return to fasting values until 8 or 9 hours after the meal (Figure 2). The most noticeable difference between the two groups was the lower triglyceride concentrations in the late postprandial period in the Southern European

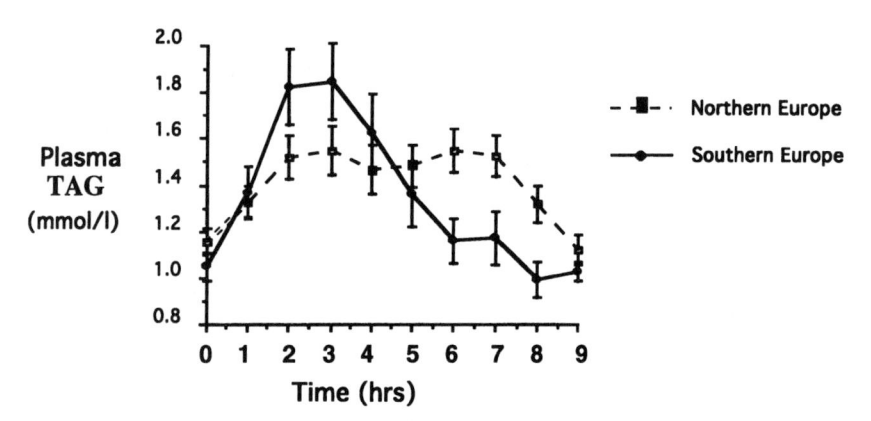

Williams et al 1995

Figure 2 *Postprandial plasma triacylglycerol (TAG) responses to standard mixed test meals containing 40 g fat in young male subjects from Southern (n=30) and Northern Europe (n=30)*

group. Whether these findings have any significance to differences in risk of CHD between these two populations is not known, but they are consistent with the observation that elevations in late postprandial triglyceride values, between and 6 and 10 hours after a meal, have been identified as a significant risk factor for CHD in a number of case control studies,[11,45] and in a study of the offspring of subjects with CHD.[12]

5 Summary

The atherogenic lipoprotein phenotype may be a common lipid abnormality in highly developed Western societies, reflecting an imbalance in lipid and energy metabolism in an increasingly sedentary society. It is linked to increased risk of CHD despite the fact that elevated cholesterol levels are not observed in most subjects. The primary defect appears to be an abnormality in triglyceride metabolism with an elevated triglyceride response to a fat-containing meal (owing either to excessive intake, overproduction or impaired clearance of triglyceride) being a major feature in most subjects. With increasing recognition of this common disorder there is likely to be greater emphasis on the need to modulate triglyceride levels by dietary and other lifestyle changes. At present our ability to advise on fatty acid intakes likely to protect against the development of an elevated triglyceride response is limited by the paucity of studies which have investigated postprandial, as opposed to fasting, lipid response following dietary fatty acid modification.

It is clear that the nature of the fatty acids fed in single meals and in background diets is important in determining the lipaemic response to a fat-containing meal. Present evidence suggests that postprandial responses to meals containing PUFA are lower than when SFA or MUFA are at high concentrations in the meal. Notably attenuated responses are seen when meals rich in LC $n-3$ PUFA are fed, but the levels used in most studies are far higher than would be present in meals commonly consumed by most human societies. The nature of the fat fed in the background diet also influences the postprandial response to a single standard meal, with the LC $n-3$ PUFA showing the greatest ability to modify lipaemic response. These effects occur at levels of intake in the diet which are observed in some societies such as the Greenland Eskimo, although they are considerably higher than those found in most Western diets.

Future work should concentrate on longer term studies using realistic intakes of LC $n-3$ PUFA or their precursor α-linolenic acid. More information is also required on the mechanism underlying the hypotriglyceridaemic effects of these fatty acids.

References

1. Committee on Medical Aspects of Food Policy, Report on Health and Social Subjects No. 46, HMSO, London, 1994.

2. Department of Health, Report on Health and Social Subjects No. 41, HMSO, London, 1991.
3. J. S. Cohn, *Curr. Opin. Lipid*, 1994, **5**, 185.
4. R.J. Havel, *Curr. Opin. Lipid.*, 1994, **5**, 102.
5. H.N. Hodis & W.J. Mack. *Curr. Opin. Lipidol.*, 1995, **6**, 209.
6. F. Karpe, G. Steiner, K. Uffelman, T. Olivecrona, & A. Hamsten, *Atherosclerosis*, 1994, **106**, 83.
7. F. Karpe & A. Hamsten, *Curr. Opin. Lipid.*, 1995, **6**, 123.
8. J.S. Cohn, E.J. Johnson, J.S. Millar, S.D. Cohn, R.W. Milne, Y.L. Marcel, R.M. Russell & E.J. Schaefer, *J. Lipid Res.*, 1993, **34**, 2033.
9. B.O. Schneeman, L. Kotite, K.M. Todd & P. Havel, *Proc. Natl. Acad. Sci. USA*, 1993, **90**, 2069.
10. J.L. Potts, R.M. Fisher, S.M. Humphreys, S.W. Coppack, G.F. Gibbons & K.N. Frayn, *Clin. Sci.*, 1991, **81**, 621.
11. J.R. Patsch, G. Miesenbock, T. Hopferwieser, V. Muhlberger, E. Knapp, & J.K. Dunn, *Arterioscler. Thromb.*, 1992, **12**, 1336.
12. C.S.P.M. Uiterwaal, D.E. Grobbee, J.C.M. Witteman, W-A. H.J. van Stiphout, X.H. Krauss, L.M. Havekes, A.M. de Bruin, A. van Tol, & A. Hofman, *Ann. Intern. Med.*, 1994, **121**, 576.
13. J.R. Patsch, *Atherosclerosis*, 1994, **110**, S22.
14. S. Sethi, M. Gibney, & C.M. Williams, *Nutr. Res. Rev.*, 1993, **6**, 161.
15. W.R. Hazzard & E.L. Bierman, *Metabolism*, 1976, **25**, 777.
16. F. Nigon, P. Lesnik, M. Rouis, & M.J. Chapman *J. Lipid Res.*, 1991, **32**, 1741.
17. A. Georgopoulos, S.D. Kafonek, & I. Raikhel, *Metabolism*, 1994, **43**, 1063.
18. M. Kinoshita, E.S. Krul, & G. Schonfield, *J. Lipid Res.*, 1990, **31**, 708.
19. S. Parthasarathy, M.T. Quinn, D.C. Schwenke, T.E. Carew, & D. Steinberg, *Arterioscher. Thromb.*, 1989, **9**, 398.
20. A. Chait, R.L. Brazer, D.L. Tribble, & R.M. Krauss, *Am. J. Med.*, 1993, **94**, 350.
21. J.C.L. Mamo, & J.R.Wheeler, *Coronary Heart Dis*, 1994, **5**, 695.
22. S.D. Krasinski, J.S. Cohen, R.M. Russell & E.J. Schaefer, *Metabolism*, 1990, **39**, 357.
23. B. Foger & J.R. Patsch, *Curr. Opin. Lipidol*, 1993, **4**, 428.
24. P.H.E. Groot, B.C.J. de Boer, E. Haddeman, U.M.T. Houtsmuller, & W.C.Hulsmann, *J. Lipid Res.*, 1988, **29**, 541.
25. E. Levy, C.C. Roy, R. Goldstein, H. Bar-On, & E. Ziv, *J. Am. Coll. Nutr.*, 1991, **10**, 69.
26. T.W.A. de Bruin, C.B. Brouwer, M. van Linde-Sibenius Trip, H. Jansen, & D.W. Erkelens, *Am. J. Clin. Nutr.*, 1993, **589**, 477.
27. W.S. Harris, W.E. Connor, N. Alam, & D.R. Illingworth, *J. Lipid Res.*, 1998, **299**, 1451.
28. M.S. Weintraub, R. Zechner, A. Brown, S. Eisenberg, & J. Breslow, *J. Clin. Invest.*, 1988, **82**, 1884.
29. A. Zampelas, A. Peel, B.J. Gould, J. Wright, & C.M. Williams, *Eur. J. Clin. Nutr.*, 1994, **48**, 88.
30. A. Zampelas, J.M.E. Knapper, K.G. Jackson, C.C. Culverwell, J. Wilson, B.J. Gould, & C.M. Williams, *Atherosclerosis*, 1995, **115**, S46.
31. P.N.M. Demacker, I.G.M. Reijnen, M.B. Katan, P.M.J. Stuyt, & A.F.H. Stalenhoef, *Eur. J. Clin. Invest.*, 1991, **21**, 197.
32. N. Yahiah, C. Songhurst, & T.A.B. Sanders, *Proc. Nutr. Soc.*, 1996, **55**, 227A.

33. K.G. Jackson, J.M.E. Knapper, A. Zampelas, B.J. Gould, J.A. Lovegrove, J. Wright, & C.M. Williams, *Atherosclerosis*, 1995, **115**, S16.
34. A. Zampelas, C.C. Culverwell, J.M.E. Knapper, K. Jackson, B.J. Gould, J. Wright, & C M. Williams, *Proc. Nutr. Soc.*, 1994, **54**, 164A.
35. W.S. Harris, & W.E. Connor, *Trans. Am. Physiol. Assoc.*, 1980, **43**, 179.
36. A.J. Brown & D.C.K. Roberts, *Arterioscler. Thromb.*, 1991, **11**, 457.
37. W.S. Harris & F. Muzio, *Am. J. Clin. Nutr.*, 1993, **58**, 68.
38. W.S. Harris & S.L. Windsor, *J. Appl. Nutr.*, 1991, **43**, 5.
39. C.M. Williams, F. Moore, L. Morgan, & J. Wright, *Br. J. Nutr.*, 1992, **68**, 655.
40. J.A. Lovegrove, C.N. Brooks, M.C. Murphy, B.J. Gould, & C.M. Williams, *Br. J. Nutr.* 1999, in press.
41. A.H. Lichtenstein, L.M. Ausman, W. Carrasco, J.L. Jenner, L.J. Gualtieri, B.R. Goldin, J.M. Ordovas, & E.J. Schaefer, *Arterioscler. Thromb.*, 1993, **13**, 1533.
42. J. Gregory, K. Foster, H. Tyler, & M.Wiseman, *British Adult Diet Survey*, *HMSO*, London, 1991.
43. C.M. Williams, A. Zampelas, K.G. Jackson, B.J. Gould, J. Wright, A. Kafatos, & M. Kapsokephalou, *Atherosclerosis*, 1995, **115**, S46.
44. A. Zampelas, H. Roche, M. Kapsokefalou, J.M.E. Knapper, K.G. Jackson, E. Pentaris, M. Tornatis, C. Hatzis, M.J. Gibney, A. Kafatos, B.J. Gould, J. Wright, & C.M.Williams, *Atheroclerosis*, 1997, in press.
45. P.H.E. Groot, W.A.H.J. van Stiphout, X.H. Krauss, H. Jansen, A. van Tol, A, E. van Ramshorst, S. Chin-On, A. Hofman, S.R. Cresswell, L. & Havekes, *Arterioscler Thromb.* 1991, **11**, 653.

6
Lipids and Obesity

John H. P. Tyman

DEPARTMENT OF CHEMISTRY, INSTITUTE OF PHYSICAL AND ENVIRONMENTAL SCIENCES, BRUNEL UNIVERSITY, UXBRIDGE, MIDDLESEX UB8 3PH, UK

1 Introduction

Obesity (also termed corpulence or a state of excessive fatness) is the accumulation of body fat usually caused by the consumption of more fat than the body can use, with the result that the excess calories are stored as fat or adipose tissue. Although overweight is not necessarily obesity, particularly where muscular or large-boned individuals are concerned, a body weight 20% or more above the optimum is considered indicative of obesity. It may also be defined quantitatively as the condition when the body mass index (BMI) is > 30, and morbid obesity when the BMI is > 40. For many years it would seem that little concern was placed on weight/age considerations. Nevertheless healthcare practitioners, government health departments and medical authorities have become united more recently in the view that one of the major challenges to human health is the control of bodyweight. Throughout Western society there has been a doubling of the number of obese people in the past decade.[1] Obesity in turn is contributory to heart disease, high blood pressure, strokes and certain forms of cancer and it has been estimated that in America, for example, 300 000 die annually from diseases related to obesity. It has been considered to be the second most important preventable cause of death in the developed world but nevertheless a medical condition which will respond to pharmaceutical treatment and a modified lifestyle. Both the more sedentary nature of this latter and an increase in the consumption of fats are important changes in recent years that have contributed to a rise in obesity. Although non-drug voluntary approaches in the slimming context would manifestly seem to be the logical approach to problems of overweight, the more overriding practice of both 'having your cake and eating it' has led over the years to drug methodology stimulated by the potential of rich, lucrative awards for the pharmaceutical industry. Nevertheless in more recent years more mundane factors have been suggested as important, namely that the damage of fatty diets has been overstressed, that weight/height charts from medical sources are questionable and that some blame for instances of obesity should be directed at junk-food diets and excessive 'nibbling' eating habits. For the excessively obese, surgery to shorten the gut has been invoked as a last resort but it seems that the power of the mind over the body can sometimes negate this approach.

Lipidic substances, although essential nutritionally, are undoubtedly implicated in both the health and ill-health of human beings. Cholesterol in fats is involved in atherosclerosis while lipids, generally of the triacylglycerol group, have an intrinsic involvement in obesity or adiposity. Although it has been generally thought that the incidence of strokes is related to the intake of fats, recent studies in the western populations have now concluded that there is an inverse association between the development of ischemic strokes and dietary fat intake.[2] It is becoming appreciated that the increase in deaths due to cardiovascular disease formerly attributed to dietary factors should now be corrected for the influence of smoking in leading to this disorder.[3]

This review will briefly touch upon the historical development of antiatherosclerotic drugs and more particularly will be devoted to the numerous approaches which in recent and past years have been studied in both understanding and in combating obesity. These have included drug methodologies for the reduction of fat digestion, for increasing the amount of fat metabolised, for the reduction of appetite and for the inhibition of fat biosynthesis. In this last category the finding of certain natural products which can restrict the activity of 'fat-synthesising' enzymes appears to be a specific targeting approach. An appendix lists the synthetic methods used for many of the compounds discussed.

2 Antiatherosclerotic Drugs

The formation of atherosclerotic plaques represents one of the main processes in cardiovascular death which includes coronary heart disease.[4] Atherosclerotic plaques commencing as fatty streaks and progressing to plaques containing cholesterol, once formed, can be the region for thrombus formation and blood platelet deposition with ultimate calcification. Formations such as these in arterial walls cause reductions in blood flow and, if present in arteries delivering blood to cardiac muscle and to the brain, can give rise to grave health problems. Thus reduction in serum cholesterol can aid the reduction of atherosclerosis and indeed it has been instanced[5] that for every 1% reduction in serum cholesterol a 2% reduction in adverse effects of coronary heart disease follows.

Cholesterol-lowering and Lipid-lowering Agents

Mevinolin and its 6-demethyl derivative. It is known that 50% or more of body cholesterol results from *de novo* synthesis.[6] Nonetheless, although a logical contributory step in the reduction of serum cholesterol is the voluntary diminution of dietary cholesterol, there has been considerable activity in pharmaceutical research to locate drugs which can influence the composition of serum lipids. Work[7] on *Penicillium citrinum* led to the isolation of a potent inhibitor of cholesterol biosynthesised from acetate which was identical to an antifungal compound, (+)-compactin (1, R = H), isolated[8] from *Penicillium brevicompactum*, 6-demethylmevinolin, which has been synthesised.[9] A more active compound in the control of cholesterol metabolism is the 6-methyl derivative of compactin, namely mevinolin (also known as lovastatin) and this compound was first described as a cholesterol-lowering agent in 1980. It operates through inhibition of β-hydroxy-β-methylglutarate-coenzyme A reductase, acyl CoA-acyltransferase and interference with low-density lipoprotein (LDL). Mevinolin, which has the tetra-substituted decalindiene structure shown (1, R = Me), is a natural product isolated from the fungus *Aspergillus terreus*.[10] It has the function of a pro-drug which is actually hydrolysed to the active β-hydroxy form. In the original clinical studies, cholesterol levels of 150–300 mg/dL in serum were decreased by 25% with an intake of lovastatin at 15mg/day during one week.

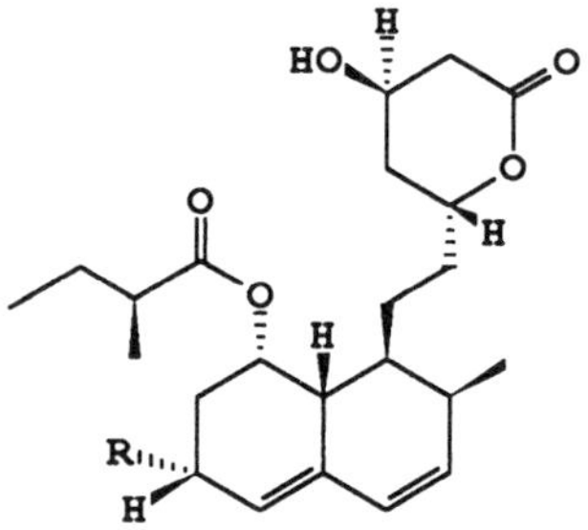

(1, R = Me) Mevinolin

Colestipol. The anion exchange polymers colestipol[11] [a derivative of diethylenetriamine $(NH_2CH_2CH_2)_2NH$ and 1-chloro-2,3-epoxypropane] and cholestyramine[12] (2) (a styryl–divinylbenzene copolymer containing quaternary ammonium groups) have the ability to bind bile acids. Since these are biosynthesised from cholesterol these two drugs can influence cholesterol metabolism and thus exert an antiatherosclerotic effect through the formation of bile acid complexes. The resulting effect is the lowering of low density lipoprotein-cholesterol (LDL cholesterol) and serum cholesterol by as much as 50% with long term treatment.[13] No absorption of the drugs in the body occurs.

Fibric acid derivatives. The alkyl fibrates are another group of well-tried compounds in the control of blood lipids. Four compounds which have been synthesised are fenofibrate[14] (3), a derivative of 4-chlorobenzophenone, gemfibrozil[15] (4), an ether of 2,5-dimethylphenol, clofibrate[16] (5), an ether of 4-chlorophenol, and bezafibrate[17] (6) derived from 4-chlorobenzoic acid. From studies of their action the compounds in this class appear to control

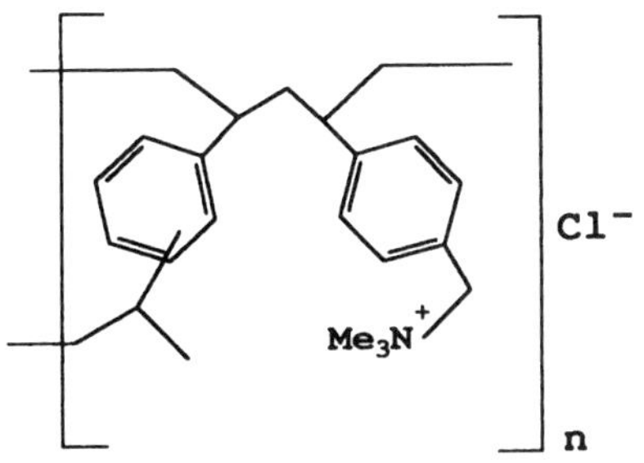

(2) Cholestyramine resin

serum lipids through the regulation of lipoprotein lipase activity, although there seem to be multiple mechanisms occurring in the lowering of serum cholesterol including an influence on lipoprotein synthesis, the control of key enzymes and of receptors concerned with lipid metabolism. Generally this group of drugs has been prescribed for certain cases of hyperlipidemia and is considered effective in reducing plasma triglycerides and LDL cholesterol. It is of interest that nicotinic acid (pyridine-3-carboxylic acid), the parent compound of nicotinamide, can in heavy doses function in reducing serum triglycerides through its action in decreasing the formation of very low density lipoprotein. Perhaps it is conjectural to infer that the lack of obesity in heavy cigarette smokers may be associated with high nicotine absorption and thence biologically produced nicotinic acid.

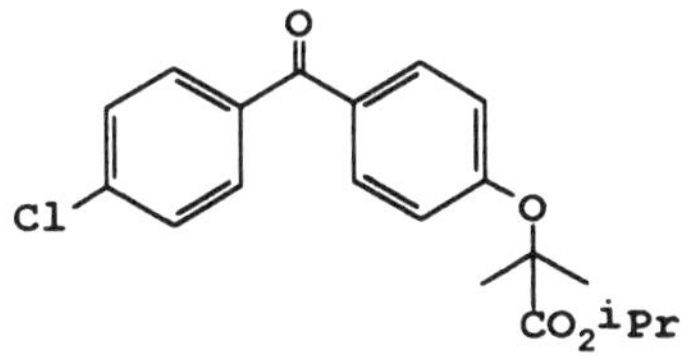

(3) Fenofibrate

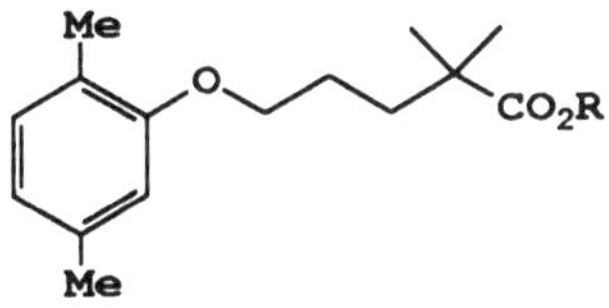

(4) Gemfibrozil

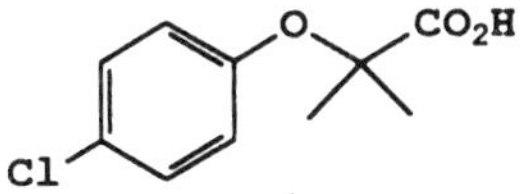

(5) Clofibrate

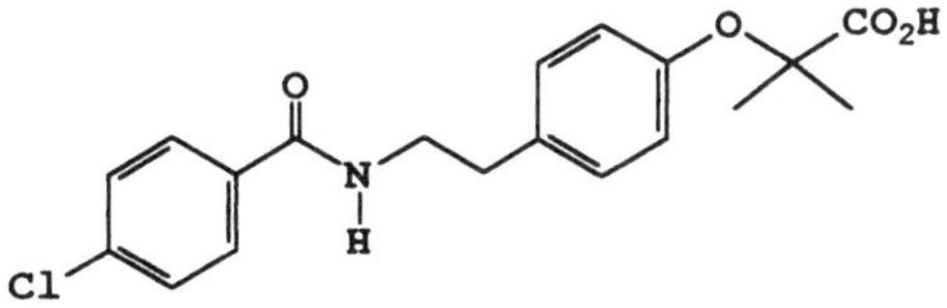

(6) Bezafibrate

Adrenoceptor antagonists. From much clinical data on the lipid composition of patients undergoing therapy for other cardiovascular diseases it is known that several groups of compounds can serve incidentally in an antiatherosclerotic capacity. Thus β-blockers (adrenoceptor antagonists) can slow the formation of plaques.[18] The accumulation of calcium compounds is considered contributory to complex processes leading to atherosclerosis and accordingly calcium channel blockers such as verapamil[19] (7) (isoptin, 5-[N-(3,4-

dimethoxyphenyl)methylamino]-2-(3,4-dimethoxyphenyl)-2-isopropylvaleronitrile) and nifedipine[20] (8) have been demonstrated in animal experiments to stimulate the low density lipoprotein receptor and thereby potentially enhance lipid metabolism in the arterial wall.[21]

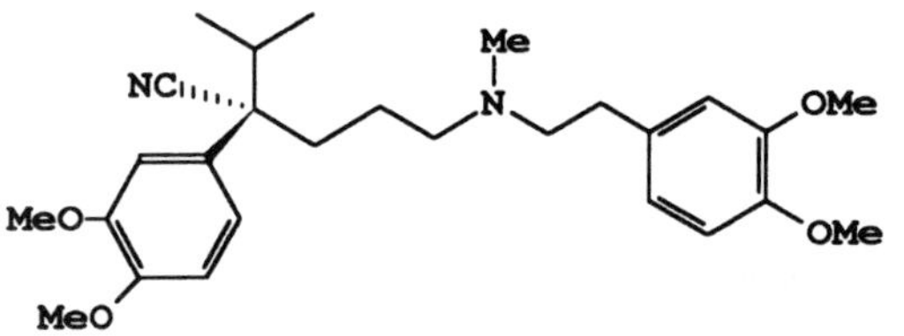

(7) Verapamil

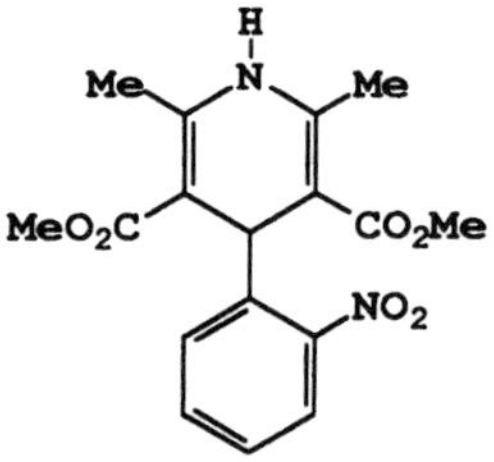

(8) Nifedipine

Antioxidants. These compounds are of inestimable efficacy in the human body and perhaps it is not altogether surprising that the compound probucol[22] (9), a derivative of 2,6-di-*t*-butylphenol, has been shown to be effective in lowering low density lipoprotein cholesterol through prevention of lipidic oxidation.

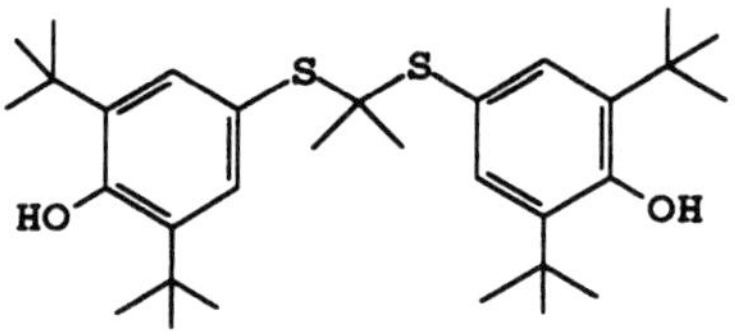

(9) Probucol

3 Drugs for the Control of Fat Absorption

Progress in the control of fat absorption has been made in the development[23] by Hoffmann La Roche and others[24] of the synthetic compound orlistat (10), the tetrahydro derivative of lipstatin (11), a natural product from *Streptomyces toxytricini*. The former is an inhibitor of pancreatic lipase, the enzyme which breaks down triacylglycerols in the small intestine. By reducing this by one tenth of the daily calorie intake (30 g/day), fat absorption into the blood is lowered and thence, through removal of this proportion of energy source, body weight should follow suit.[25] Apparently the absorption of essential fat-soluble vitamins is not affected although the replenishment of diminished fat reserves by carbohydrate-derived fats is a possibility. The results of a 6-month study of the efficacy and tolerability of orlistat have been described.[26] Under the name Xenical, it has recently been made available only on prescription in the UK. However, less desirable personal features appear to accompany its use and it has been suggested that loss of fat-soluble vitamins in unadsorbed fat could result in avitaminosis.

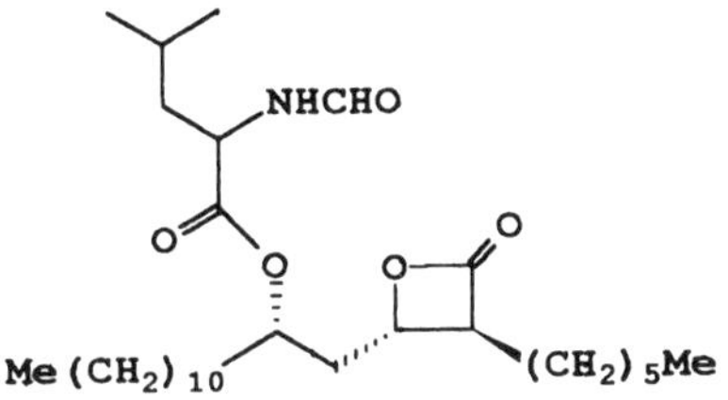

(10) Orlistat

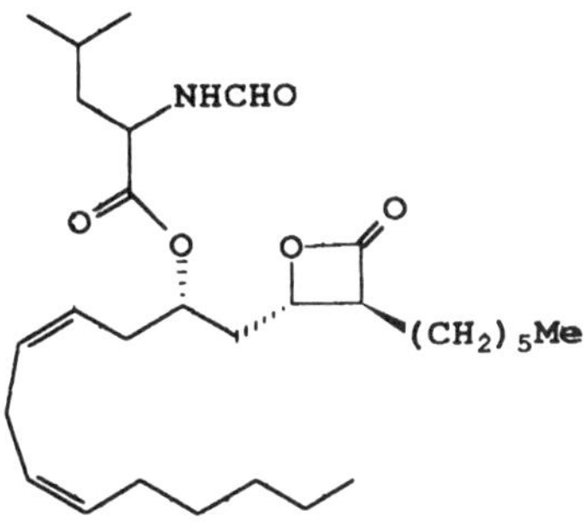

(11) Lipstatin

4 Thermogenic Drugs for Increasing the Metabolism of Certain Types of Fat

It has been found[27] that brown adipose tissue (BAT), the colour of which is due to iron-based cytochromes, contains cells capable of being activated by organic compounds and thence of oxidising fats with generation of body heat, a metabolic process termed thermogenesis. Although practical body weight reduction through this pathway has been mediated with compounds such as *pseudo*-norephedrine, side effects occur and only recent pharmacological studies with rodents have established the existence of a β_3 receptor in BAT cells sensitive to treatment with adrenaline and noradrenaline, leading to thermogenic response and thence gradual reduction in body fat.[28]

Now several more selective β_3 agonists have been produced, including the unsymmetrical bisphenylethylamines (12) (BRL 35135, Beecham[29]) and (13) (CL 316243, American Home Products[30]), a trisphenylethylamine relative (14) (Ro 40-2148, Hoffmann La Roche[31]) and the substituted 1,4-dioxybenzene compound (15) (ZD 7114, Zeneca[32]), which is [(*S*)-4-(2-(hydroxy-3-phenoxypropyl)amino)ethoxy]-*N*-(2-methoxyethyl)phenoxyacetamide. In animal studies[33] these compounds enhanced the metabolic thermogenesis of fats without cardiovascular effects. Thus with (13), CL 316243, in the case of adult beagles, daily treatment over 5–7 weeks led to decreased weight and girth compared to controls, while (15), ZD 7114, also showed a lipomobilising effect. Other investigations[34] showed that treatment with ZD 7114 (15) significantly reduced weight gain and activated brown adipose tissue. Complex situations appear to exist with these compoinds and the prospect of an appropriate compound for human application has still to be demonstrated.

5 Compounds for the Control of Appetite

In the animal world there has been a considerable compilation of information on anti-feedants, although with humans the related area of appetite control

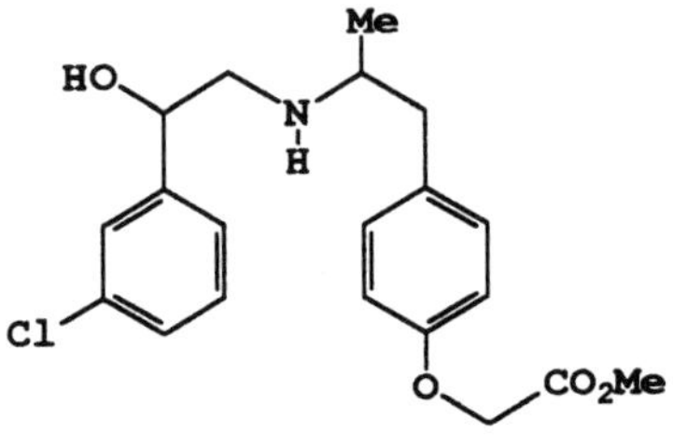

(12) BRL 35135

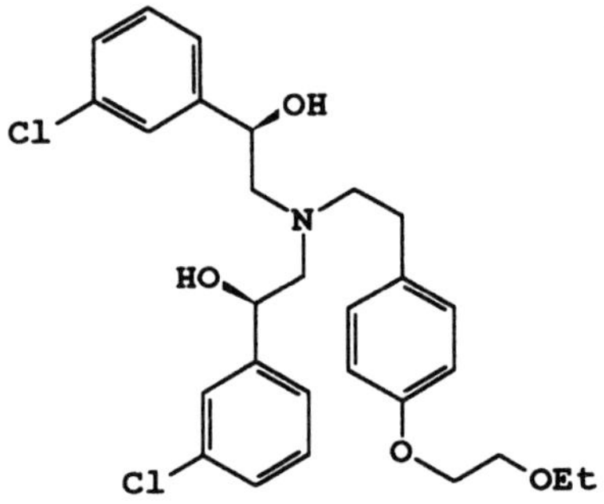

(13) Ro 40-2148

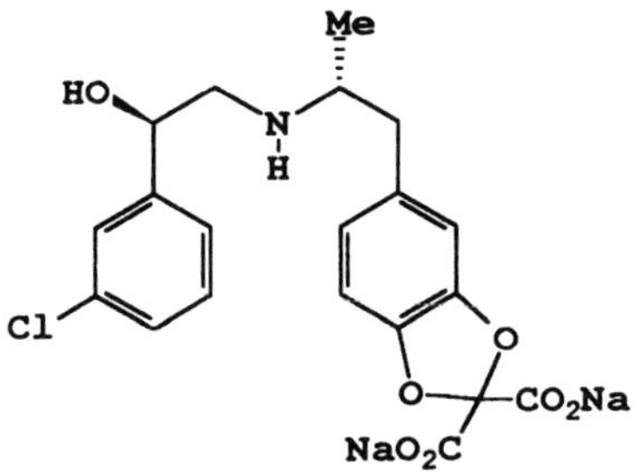

(14) CL316243

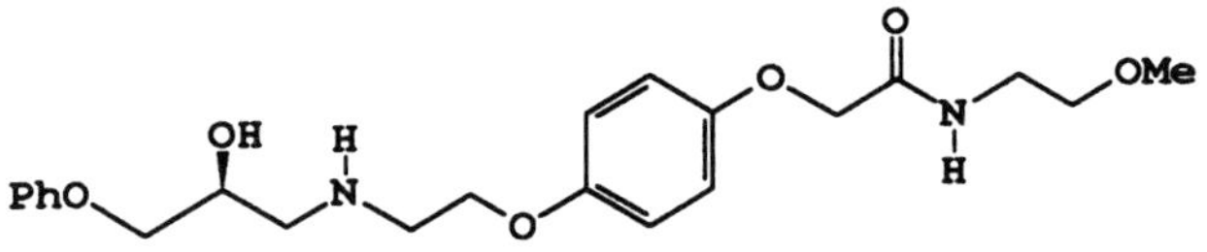

(15) ZD 7114

has been studied in a different way. Anti-eating drugs or anorectics have been known for some time, and more recently research on their mode of action has been intensified.

Probably inclusive in this category of compound is the sucrose polyester 'olestra' which tastes like fat but passes through the body without being digested. Although a great success in the USA, doubts have been raised about its significant side-effects, apparent from certain trials,[35] and in unpleasant features accompanying its use.

Chemical Studies

Although one of the first significant approaches in the control of appetite was based on the use of amphetamines, their prescriptive use was prohibitive since they are classed as restricted drugs due to their addictive properties, quite apart from stimulant and mood-influencing side effects. Amphetamines act by increasing the concentration of neurotransmitters such as dopamine [2-(3,4-

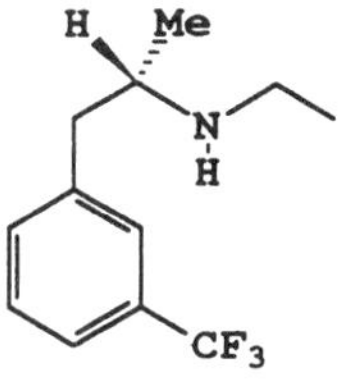

(16) (S)-Fenfluramine

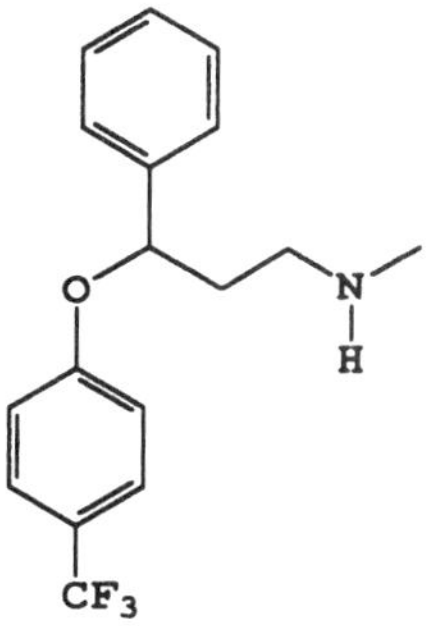

(17) Fluoxetine

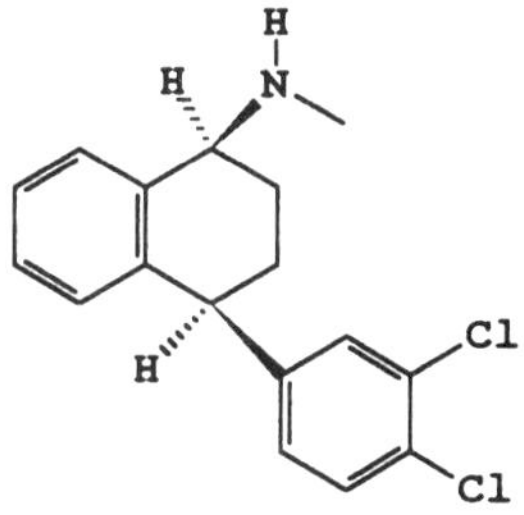

(18) Sertraline

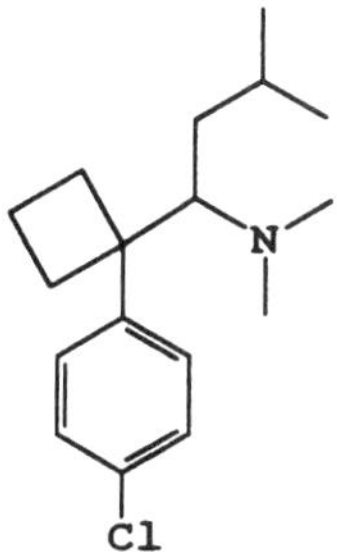

(19) Sibutramine

dihydroxyphenyl)ethylamine] and noradrenaline [α-(aminomethyl)-3,4-dihydroxybenzyl alcohol] which have been associated in this connection with food intake. By contrast, serotonin [5-hydroxytryptamine, 3-(β-aminoethyl)-6-hydroxyindole] has been linked to the completion of this stage and thus to satiety. The structures of amphetamines have been emulated in a range of drugs produced by Servier Laboratories. Thus (S)-fenfluramine, namely dexfenfluramine[36] (16), has the effect of suppressing the appetite through serotogenic action.[37] Nevertheless side effects have restricted its general prescriptive usefulness, and although it had been approved in the USA by the Food and Drug Administration it has been voluntarily withdrawn[38] by its manufacturer on account of its non-specificity and side effects experienced in its use in combination with antidepressant drugs ('fen-phen health scare'). Other antidepressant-type compounds which regulate serotonin levels, such as fluoxetine[39] (17) (prozac, Dista Products) and sertraline[40] (18) (lustral, Invicta Pharmaceuticals), have been shown in animal studies[41] to have weight-reducing properties, although less clearly so with humans. Sibutramine,[42] a satiety drug (19) (Knoll Ltd), has been developed as a compound which results in a reduction in food intake through the action of increased concentrations of noradrenaline and of serotonin. Following trials it is now available for use in the USA.

Amongst the natural compounds studied as satiety agents in the 1970s was the gastrointestinal hormone cholecystokinin (CCK). A series of peptides are involved in its circulation and its peripheral effects comprise contraction of the gall bladder, stimulation of enzyme secretion from the pancreas and the slowing of gastric emptying. As a satiety agent it triggers receptors in the stomach when full to signal the brain. From initial work by Merck laboratories on non-peptide compounds mimicking CCK, a novel benzodiapine (20) was developed[43] which possessed some CCK-A agonist activity, although it was not orally effective in animal feeding studies. Structural variants of this compound have been prepared in which the amide portion of the molecule has been altered and the indazole (21, R^1 = OMe, R^2 = H, X = N and Y = NH) was found to be the most effective and capable, when given orally, of reducing food intake by 40% in doses of 10 μmol/kg after 3 hours.

(20)   (21)

On an empirical level it has been claimed[44] that inhibiting the digestive system can be achieved with a mixture of L-phenylalanine, tryptophan, methionine and valine together with oleic acid, D-dextrose and sodium taurocholate and that the high calorific value of high energy foods can be reduced with a consequential potential weight loss.

Biochemical and Genetic Work

Fundamental work has shown that while non-peptide compounds like noradrenaline are significant in their role, a neuropeptide Y (or NPY) synthesised in the hypothalamus is a key influence in several physiological processes[45] through its ability to interact at several different types of receptor termed Y1, Y2, Y3 and Y4 subtypes. Indeed it had been suspected that another novel receptor[46] called Y5 with a specific profle is influential in the mediating role of NPY in the hunger/satiety cycle. Thus through a proposed genetic/biotechnological approach it is now possible in principle to develop oral drugs for human use with the potential to act as antagonists to the Y5 receptor and to result in reduced food intake.

As well as great interest in NPY there has been much work in recent years on the lipostatic hormone leptin, a 16-kDa protein comprising 167 amino acids, which was discovered in fat cells during the course of endeavours to find the abnormal gene involved in producing massive obesity in the *ob/ob* mouse.[47] It was shown that this animal's obesity was due to the lack of the normal gene responsible for producing leptin and indeed when normal and *ob/ob* mice were injected with the hormone both groups underwent a reduction in body weight through reduced food intake and through energy expenditure by thermogenesis.[48,49] Although it was postulated that obesity was attributable to a failure to produce adequate amounts of leptin, subsequent studies by others[50] indicated that, in proportion to their weight, obese persons did actually produce leptin. This apparent anomaly has been approached by detailed studies on the pathway by which leptin acts. It is now established that in the brain the leptin is specifically transported to receptors in the hypothalamus at the base of the brain, but in the obese the amount of leptin carried to the brain is not proportional to that in the blood.[51,52] Several explanations have been advanced, namely that either the transport system is already saturated or that it has been attenuated or inactivated to leptin. There is little doubt that leptin is implicated in the regulation of food intake and energy expenditure, and the whole-body energy balance in rodents and humans and its biology has been reviewed.[53] Low levels of leptin increase hunger pangs, reduce body weight, lower metabolism and are found in people who have dieted, while high levels are found in fatter people who would appear to become insensitive to its normal signalling action in the hypothalamus. The excessively obese appear to lack a specific gene.

It has been found that there are links between the leptin and the NPY systems and these may be abnormal or disturbed in the obese.[54,55] It has been concluded that pharmaceuticals (or perhaps natural products) may have a role in addressing this situation.[56]

More recently two new hormones termed orexin-A and orexin-B have been found which are released by nerve cells in the brain and which are known to play an important part in appetite control.[57] These compounds are considered to be identical to the hypocretins[58] which have been described as hypothalamus-specific peptides with neuroexcitatory activity.

6 Inhibition of Fat Biosynthesis

Since obesity can be defined as the excessive accumulation of simple lipids such as cholesterol and triacylglycerols in adipose tissue, the regulation of their formation could be regarded as an important approach in its control. The synthesis of cholesterol can be regulated by substances such as compactin through the enzyme involved, namely hydroxymethylglutaryl CoA reductase. By contrast there is no similar effective inhibitor of glycerol-3-phosphate dehydrogenase (GPDH), a key enzyme in the synthesis of triacylglycerols. It has been reported that the activity of this enzyme was increased several hundred-fold when 3T3-L1 cells, a clone of Swiss/3T3 fibroblasts, were transformed to adipose cells under certain conditions of culture.[59] Several cell components, for example ATP, ADP, NAD[60,61] and fatty acids,[62] have been demonstrated to inhibit GPDH weakly, but more recently the phenolic lipids 5-pentadecylresorcinol (1,3-dihydroxy-5-pentadecylbenzene), (15:0)-cardol (22), isolated earlier[63] from *Anacardium occidentale* but ssubsequently from *Streptomyces cyaneus* obtained[64] from a Japanese soil, together with its iso analogue, 5-isopentadecylresorcinol) (23), and termed adipostatins A and B respectively, have been reported to have a much higher level of activity.[64]

(22) (15:0)-Cardol, Adipostatin A

(23) 5-isoPentadecylresorcinol Adipostatin B

Related phenolic lipids, the (13:0), (15:1), (17:1) and (17:2) anacardic acids (24), (25), (26) and (27) respectively, from *Ginkgo biloba* have also been described[65] as having inhibitory properties towards GPDH. The 5-alkenyl- and 5-alkadienylresorcinols isolated[66] from *Cereale secale* (rye) have related antienzyme activity. The activity of the phenolic lipids (22) and (23) in terms of the IC_{50} values, were 4.1 and 4.5 μg/cm^3 compared with 9.4, 4.0, 4.6 and 2.4 μg/cm^3 for (24), (25), (26) and (27) respectively. Thus the (17:2) compound appeared to be the most active and to be specific in its action. Compounds (22) and (23) were found[64] to have a ten-fold activity compared with the monohydroxyphenol, 3-pentadecyphenol.

(24) (13:0)-Anacardic acid

(25) (15:1)-Anacardic acid
Ginkgolic acid

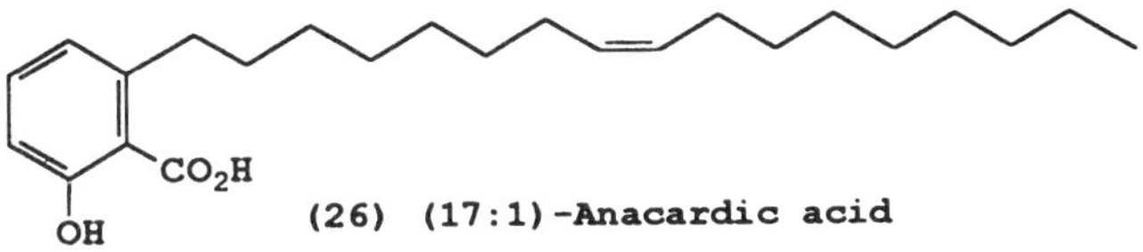

(26) (17:1)-Anacardic acid

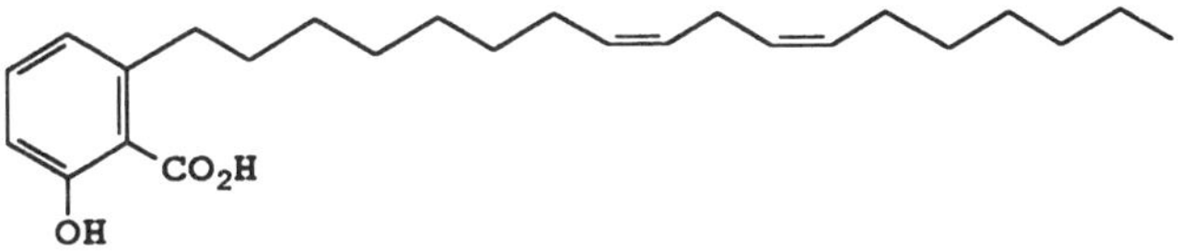

(27) (17:2)-Anacardic acid

Appendix

In this appendix the syntheses of a number of the compounds discussed in the preceding text are described.

Synthesis of Cholesterol-lowering and Lipid-lowering Compounds

(+)-Compactin (1a), 6-Demethylmevinolin, 6-Demethyl-(1). The synthesis of (+)-compactin has been achieved[9] starting from the readily available *trans*-dione A. Reduction of this with *Aureobasidium pullulans* NRRL Y-1260 gave a 33% yield of the required diol, the sodium salt of which from NaH and DMSO was benzylated to give the dibenzyl ether. Reaction of the double bond in this with phenylselenyl bromide in acetic acid gave a phenylseleno acetoxy compound and saponification, oxidation to the selenoxide and elimination of the selenoxy group afforded the allylic alcohol and thence by Jones oxidation the α,β-unsaturated ketone shown (Scheme 1). This with a lithiocuprate reactant in THF at $-20\,°C$ gave a mixture of methylols which were mesylated and subjected to elimination to yield the corresponding methylene compound. Hydrogenation in the presence of pyridine to suppress isomerisation afforded the axial methyl compound. The tolylhydrazide with lithium diethylamide in THF at $-78°$ to $0\,°C$ gave the corresponding alkene which was debenzylated to the diol with lithium in ammonia. With an excess of (S)-2-methylbutyric anhydride the diester resulted and hydrolysis gave a predominant amount of the less hindered derivative which was mesylated. Elimination treatment afforded the alkene and removal of the protecting group R (from ethyl vinyl ether) gave the alcohol which upon oxidation gave the corresponding aldehyde. This, with the dianion of methyl acetoacetate, gave a mixture of diastereoisomeric alcohols which owing to their inseparability were reduced to two pairs of β,δ-dihydroxy esters. The less polar pair of these was separated chromatographically and lactonisation afforded two β-hydroxy lactones from which by HPLC separation the desired product was isolated and found to be identical with natural (+)-compactin. The authors reported that the synthesis could be adapted to that of mevinolin itself.

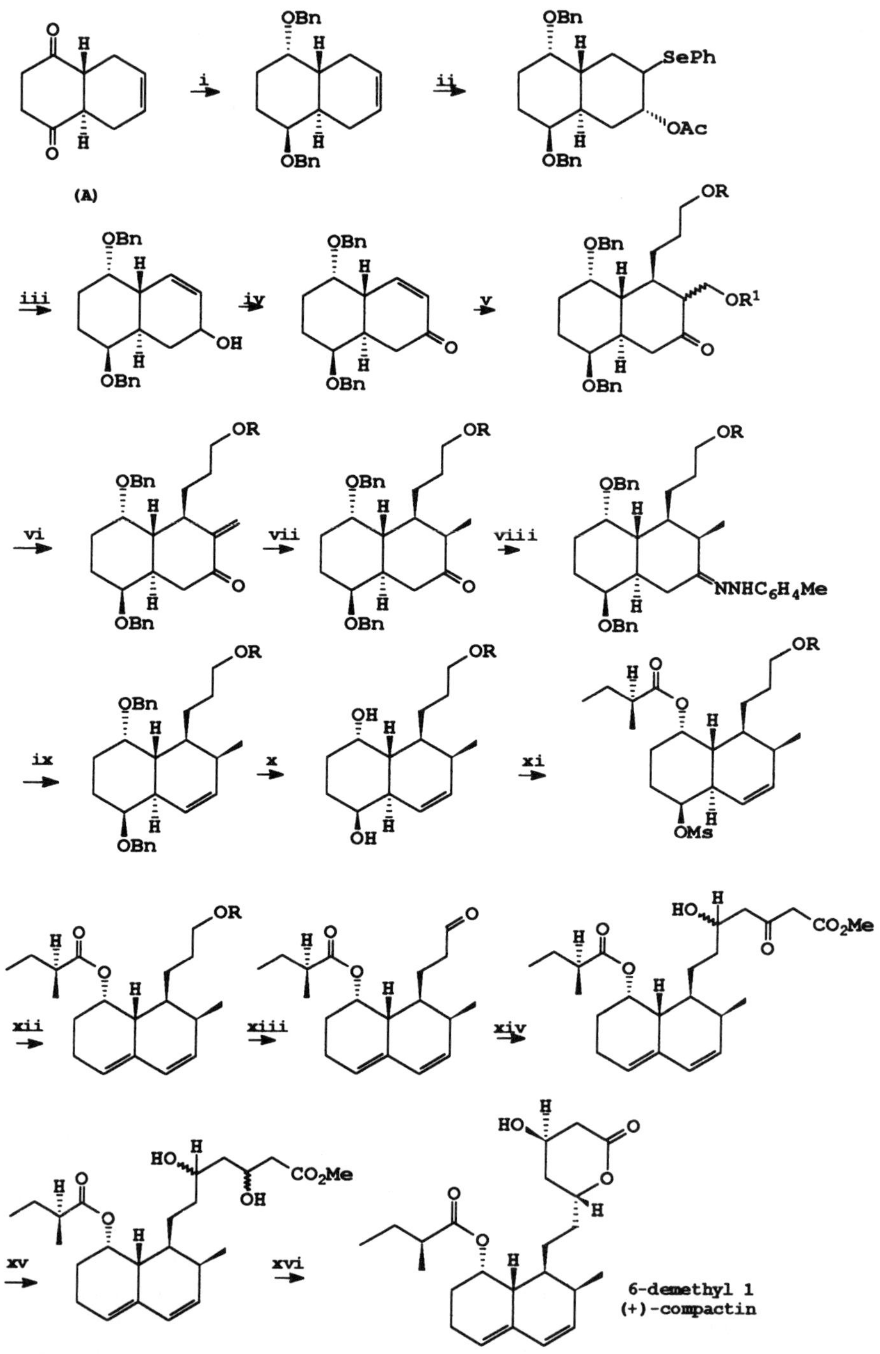

O
H
O
(A)
i
OBn
OBn
ii
SePh
OAc
iii
OH
iv
v
OR
OR¹
vi
vii
viii
NNHC6H4Me
ix
x
OH
xi
OMs
xii
xiii
xiv
HO
CO2Me
xv
OH
xvi
6-demethyl 1
(+)-compactin

Scheme 1

Scheme 1 *Synthesis of (+)-compactin*

Reactants: (i) Redn. with *Aureobasidium pullutans*; 2NaH, DMSO, BnCl; (ii) PhSeBr, AcOH, KOAc, 25 °C, (iii) KOH/MeOH-Et$_2$O; H$_2$O$_2$, THF, 25 °C; Δ, 55 °C; (iv) Jones oxidn.; (v) RO(CH$_2$)$_3$CuLi(SPh), THF, −20 °C (R = CH(Me)OEt; CH$_2$O(g), THF, −78 °C; MeSO$_2$Cl/Et$_3$N/CH$_2$Cl$_2$, (−5 °C (R^1 = Ms); (vi) DBU/PhH, Δ, 25 °C; (vii) Pd–C, H$_2$, Py; (viii) MeC$_6$H$_4$NHNH$_2$, PhH, 25 °C; (ix) Xs LDA,THF, −78 °C; (x) Li/NH$_3$ (anhyd.); (xi) Xs (S)-[MeCH$_2$CH(Me)CO]$_2$O, Py /DMAP, 25 °C; Xs KOH/EtOH, 25 °C; MeSO$_2$Cl, Py, 0 °C; (xii) DBU, Py, 115 °C; (xiii) AcOH, THF, H$_2$O; PCC, CH$_2$Cl$_2$, 25 °C; (xiv) $^-$CH$_2$CO$^-$CHCO$_2$Me, THF, 0 °C; (xv) Xs Zn(BH$_4$)$_2$, 0 °C; (xvi) chromatogr., CC, SiO$_2$, Me$_2$CO–CHCl$_3$ (1:9); 4-TsOH–H$_2$O, PhH, 25 °C; HPLC of diastereoisomers.

The Extraction of Mevinolin (1, R = Me). The extraction of mevinolin from a strain of *Aspergillus terreus* has been described.[10] The culture from agar slants, involving cell propagation in one medium and product formation in a second medium, was extracted in aqueous suspension with ethyl acetate and after the addition of Supercel the mass was filtered. The combined filtrates and ethyl acetate washings were washed with water, the organic solvent layer dried with magnesium sulfate and concentrated *in vacuo*. After addition of toluene and re-evaporation, the mevinolic acid present was converted to mevinolin by refluxing with toluene, during which water was removed azeotropically. After removal of toluene, the dark oily residue was slurried in ethyl acetate–dichloromethane (30:70) with silica gel and chromatographed on silica gel and elution with ethyl acetate–dichloromethane (40:60). By HPLC, fractions rich in mevinolin were located and recrystallisations from acetonitrile and then ethyl acetate afforded the product as a nearly white solid.

Fenofibrate (Isopropyl Ester) (3). A wide variety of classical and more recent methods are applicable to this compound[14] and its isomers. Thus Friedel–Crafts reaction of anisole (or phenol) with 4-chlorobenzoyl chloride, followed by demethylation in the case of the ether, or Fries rearrangement of phenyl 4-chlorobenzoate, could provide routes to the compound 4-chloro-4'-hydroxybenzophenone (Scheme 2). Other methods have been indicated (ref. 59b, p. 196). Reaction of phenols in sodium hydroxide solution containing propanone and chloroform has been described[16] in the herbicide field as affording α-(aryloxy)isobutyric acids, which by esterification with isopropanol would in the present case give the final product.

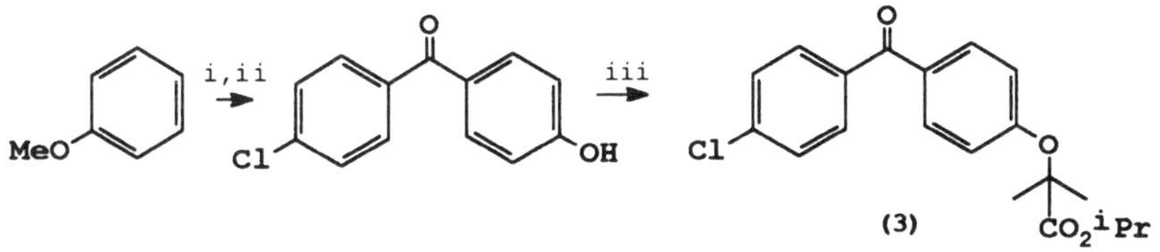

Reactants: (i) ClC$_6$H$_4$COCl, AlCl$_3$, PhNO$_2$, (ii) HBr, (iii) $^-$OH, Me$_2$CO, CHCl$_3$; iPrOH, H$_2$SO$_4$.

Scheme 2 *Synthesis of fenofibrate (3)*

Gemfibrozil (4). This compound represents one of many synthesised[15] from a number of 2,5-disubstituted phenols by reaction most systematically from a HO-protected chloro- or bromo-hydrin, conversion of this, after removal of the protective group, to an ω-chloro or bromo intermediate and alkylation of the dianion of isobutyric acid or the carbanion of an alkyl isobutyrate (Scheme 3). Gemfibrozil as Merck 4394 is listed as the free acid.

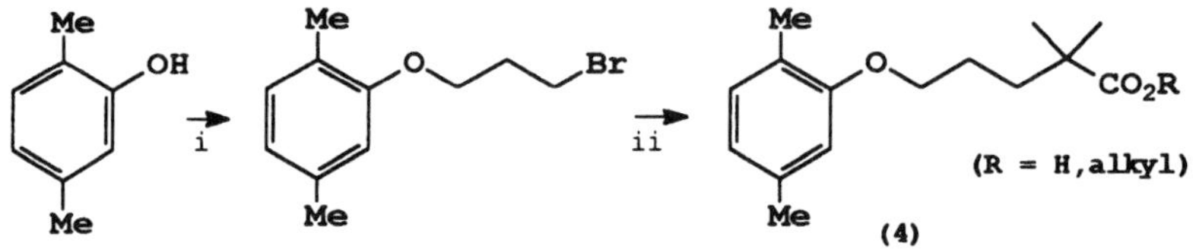

Reactants: (i) 2,5-$Me_2C_6H_3O^-$, $Cl(CH_2)_3OR$ (R = 2-tetrahydropyranyl), aprotic solvent; dil. HCl; PBr_3, Et_2O; (ii) $Me_2C^-CO_2R$ (from Me_2CHCO_2R and $LiNiPr_2$), 2,5-$Me_2C_6H_3O(CH_2)_3Br$, THF, 0 °C.

Scheme 3 *Synthesis of Gemfibrozil (4)*

Clofibrate (5). The free acid can be obtained,[16] as indicated in the case of fenofibrate, by reaction under aqueous conditions of 4-chlorophenol in sodium hydroxide solution containing propanone and chloroform in 37% yield but is accompanied in this reaction by 4-chlorophenyl orthoformate (Scheme 4). An alternative development leading to the methyl ester has been disclosed[67] (ref. 63b, p. 71) in 86% yield consisting of the reaction of 1,1,1-trichloro-2-methylpropan-2-ol in methanol to which sodium hydroxide was added. The difference from the earlier work[16] appears to be mainly in the pre-preparation of the chloro intermediate. Clofibrate as the ethyl ester is Atromid S. Clofibric acid is Merck 2437.

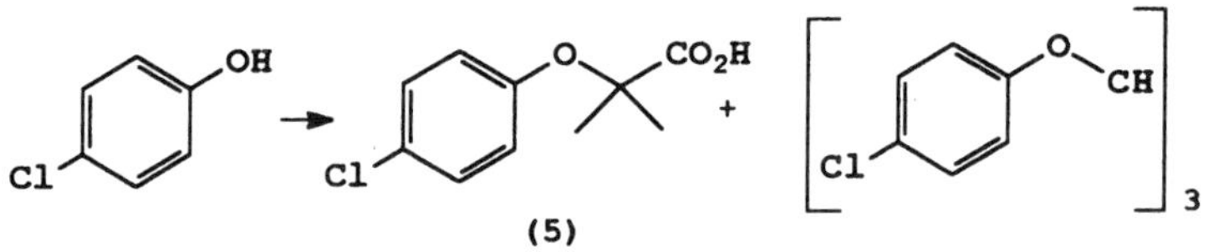

Scheme 4 *Synthesis of clofibrate (5)*

Bezafibrate (6) (free acid or ester). This 4-chlorobenzamide derivative was prepared[17] from the starting material 2-(4-hydroxyphenyl)ethylamine, itself available from a number of alternative routes. The sodium salt from hydrolysis of the mono *N*-4-chlorobenzoyl derivative (B) (obtained by hydrolysis of the corresponding *O,N*-bis(chlorobenzoyl) compound) was reacted in toluene solution with ethyl α-bromoisobutyrate and the product then hydrolysed to the acid (Scheme 5). The alternative routes[16,67] may be feasible.

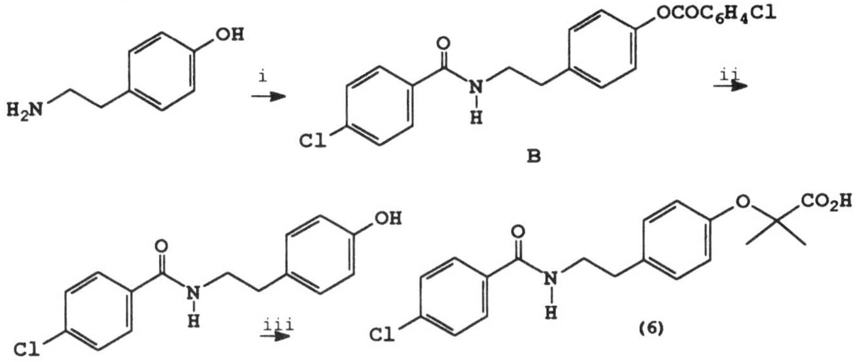

Reactants: (i) ClC$_6$H$_4$COCl, py, Δ, 100 °C; MeOH, KOH, 40–50 °C; (ii) NaOMe, MeOH, B, vac.; (iii) PhMe, Me$_2$CBrCO$_2$Et, Δ, 25 h, 80 °C; aq. KOH, HCl.

Scheme 5 *Synthesis of bezafibrate (6)*

Synthesis of Adrenoceptor Antagonists

Verapamil (7). Although the racemic form of this drug is used in the USA for cardiovascular disorders, the (2S)-(−)-enantiomer is more active and there is also interest in the (−) form of the 3,4,5-trimethoxy analogue, [(−)-gallopamil].

The synthesis of (2S)-(−)-verapamil commenced[19] with (2S)-(+)-1,2-propanediol, which was protected at the primary alcohol by tritylation and at the secondary by mesylation (Scheme 6). Prolonged reaction with the dilithium

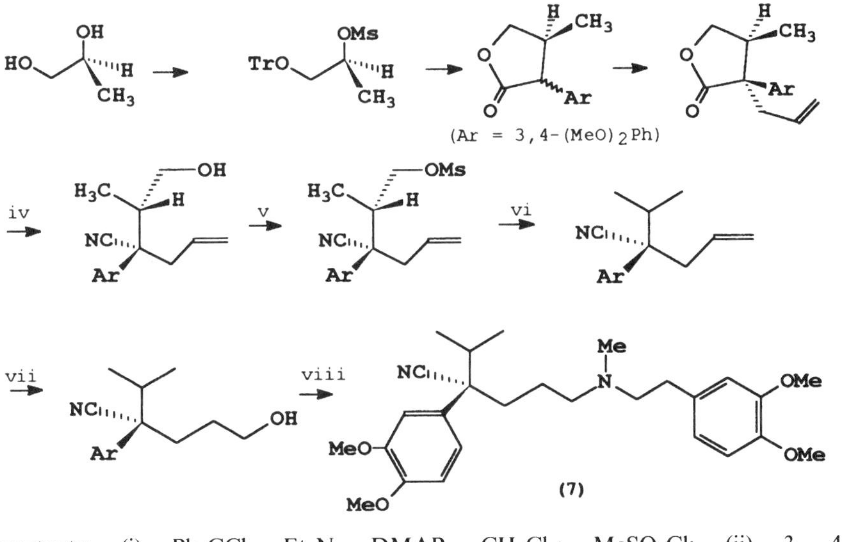

(Ar = 3,4-(MeO)$_2$Ph)

Reactants: (i) Ph$_3$CCl, Et$_3$N, DMAP, CH$_2$Cl$_2$; MeSO$_2$Cl; (ii) 3, 4-(MeO)$_2$C$_6$H$_3$CH$^-$CO$_2^-$ (from LiNiPr$_2$), THF; 4-TsOH, MeOH; (iii) NaH, THF, BrCH$_2$CH=CH$_2$; (iv) Me$_2$AlNH$_2$, Cl$_2$CHCH$_2$Cl; (v) MeSO$_2$Cl, Et$_3$N, CH$_2$CL$_2$; (vi) NaBH$_4$, *t*-BuOH, (MeOCH$_2$)$_2$; (vii) disiamylborane, THF; H$_2$O$_2$; (viii) MeSO$_2$Cl, Et$_3$N, CH$_2$Cl$_2$; 3,4-(MeO)$_2$C$_6$H$_3$CH$_2$CH$_2$NHMe, Et$_3$N, THF.

Scheme 6 *Synthesis of (2S)-(−)-verapamil*

salt of 3,4-dimethoxyphenylacetic acid in THF at ambient temperature led to displacement of mesylate and, after acidic treatment to remove the trityl group, the butyrolactone depicted was obtained as a *cis/trans* mixture. The enolate was allylated and the (2S,3S)-(−) product obtained, the ester group in which was efficiently converted to the cyano by extended reaction with dimethylaluminium amide in refluxing trichloroethane. After conversion of the hydroxymethyl to methyl by selective reduction of the derived mesylate with sodium borohydride in the presence of *tert*-butanol in order to preserve the alkene and cyano groups, hydroboration with disiamylborane in THF followed by oxidation with hydrogen peroxide afforded the primary alcohol shown. The mesylate upon treatment with *N*-methylhomoveratrylamine in THF gave the required product (2S)-(−)-verapamil, in an overall yield from the diol of 45%. The enantiomer, (2R)-(+)-verapamil was similarly obtained by starting with (2R)-(−)-1,2-propanediol. Gallopamil can be synthesised by a similar methodology.

Nifedipine (8). This compound is accessible by a two-stage classical synthesis[20] from the reaction of 2-nitrobenzaldehyde with 2 molar proportions of methyl acetoacetate in refluxing methanol solution containing ammonia (Scheme 7). In the original patent a family of related compounds was obtained and some of the nitro members were hydrogenated under pressure in ethanol containing Raney nickel to afford the corresponding amino derivatives.

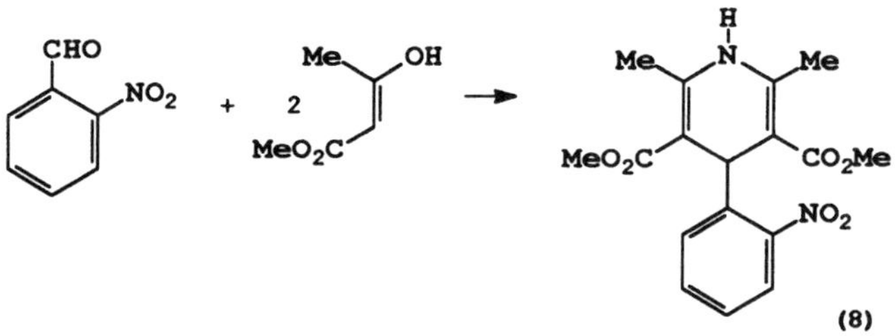

Scheme 7 *Synthesis of nifedipine*

Antioxidants

Probucol (9). This compound[22] has been synthesised from phenol by way of 2,6-di-*tert*-butylphenol, disulfurisation of which gives the disulfide shown and thence, by reduction, 4-mercapto-2,6-di-*tert*-butylphenol (Scheme 8). Reaction of this in methanolic hydrochloric acid with propanone afforded the required 4,4′-(isopropylidenedithio)bis(2,6-di-*tert*-butylphenol) in 87% yield.

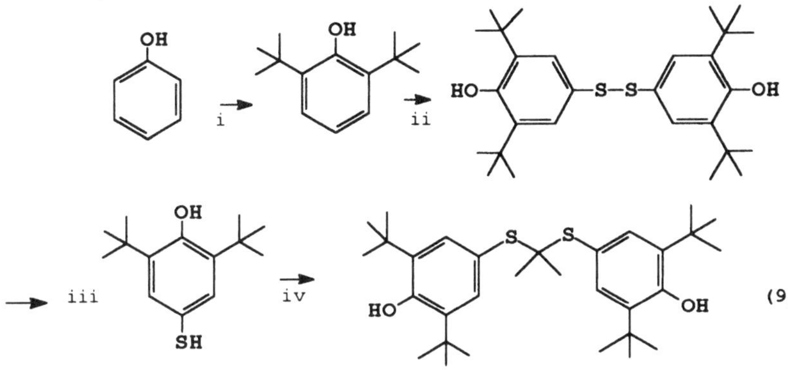

Reactants: (i) (Me)$_2$C=CH$_2$, Al catalyst; (ii) S$_2$Cl$_2$, PhMe, I$_2$, −15 °C; (iii) Zn, PhMe, H$_2$O, HC; (iv) Me$_2$CO, MeOH, HCl.

Scheme 8 *Synthesis of probucol*

Drugs for Controlling Fat Absorption

Orlistat (10) (Tetrahydrolipstatin) Although this compound is accessible from the reduction of of the naturally-occurring unsaturated analogue (11), there has been considerable activity by a number of groups in the total synthesis of the saturated compound, orlistat (10), notably in the first synthesis[23] which was achieved in remarkably few steps.

In this the available ethyl (R)-3-hydroxy-6-heptenoate, protected as the tetrahydropyranyl ether, was ozonolysed and the resultant aldehyde by the Wittig reaction gave the C$_{14}$ unsaturated ester (C) (Scheme 9). The aldehyde by reduction with diisobutylaluminium hydride upon aldol condensation with the dianion of octanoic acid afforded a mixture of stereoisomeric hydroxy acids which were lactonised to the corresponding oxetanes. Deprotection and chromatographic separation gave two *trans* and two *cis* isomers. The *trans* isomer depicted by esterification with (S)-N-formylleucine with inversion of configuration (Mitsunobu reaction) at the OH-centre gave the essential structure and tetrahydrolipstatin (orlistat) was then obtained by hydrogenation of the alkene double bond. A variation on this scheme was also described and reference made to the determination of the absolute configuration of both lipstatin and of tetrahydrolipstatin.

Subsequently, two further syntheses from the same group were described[68] in which methyl (3R)-hydroxytetradecanoate obtained by asymmetric reducion of the β-keto ester was used. 3(R)-benzyloxytetradecanal has been employed as starting material[69] and one of the most recent approaches[24] also used this intermediate as well as the allylboronation of dodecanal, giving orlistat in an overall yield of 38% by seven steps.

The structural analogue with a methylene group in place of the carbonyl group has been prepared[70] and found to be as active as orlistat when tested against porcine pancreatic lipase.

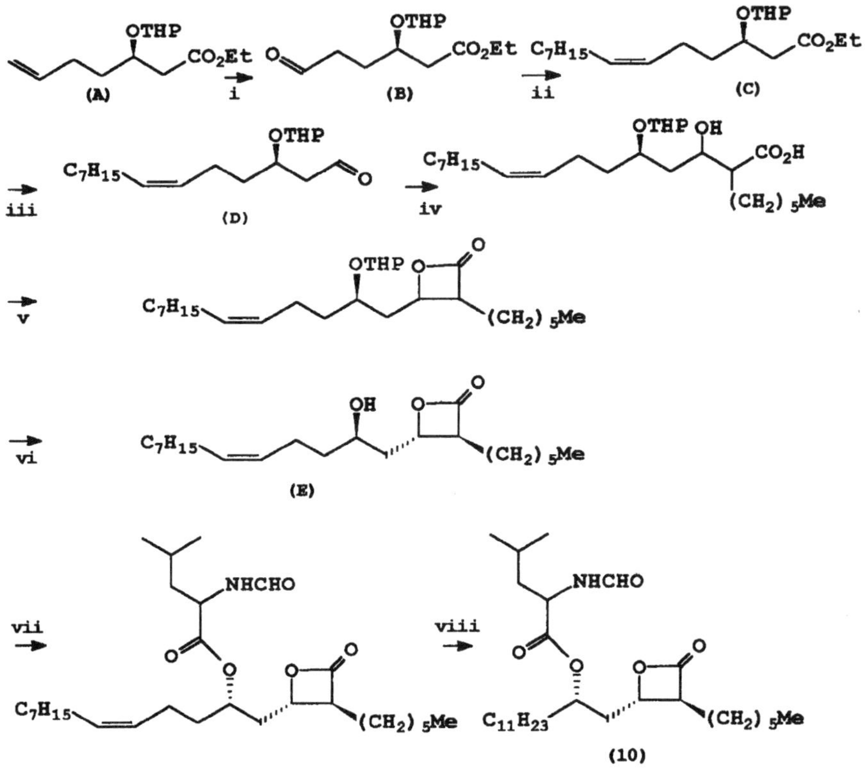

Reactants: (i) A, EtOAc, O$_3$, -78 °C; Pd-C, H$_2$, (ii) C$_8$H$_{17}$P$^+$Ph$_3$ Br, Et$_2$O THF, BuLi; B, Et$_2$O; H$_2$O; (iii) DIBAL, PhMe, C, CH$_2$Cl$_2$, -78 °C; H$_2$O; (iv) iPr$_2$NLi, THF, 0 °C, C$_7$H$_{15}$CO$_2$H; D, THF, -78 °C; aq. NH$_4$Cl; (v) Py, PhSO$_2$Cl, 0 °C; H$_2$O; (vi) EtOH, Py, 4-TSA, 50 °C; chromatography, C$_6$H$_{14}$–EtOAc, (8:2) to separate 2 more polar *trans* isomers (E and diastereoisomer) from 2 *cis* isomers, (vii) E, PPh$_3$, (S)-N-formylleucine, THF, EtO$_2$CN=NCO$_2$Et, ambient temp.; chromatography, PhMe–EtOAc, (9:1); (viii) Pd–C, H$_2$, THF; chromatography (same solvent).

Scheme 9 *Synthesis of orlistat*

Thermogenic Drugs for Increasing the Metabolism of Certain Types of Fat

The compounds (12) and (14) in this category are structurally related and are probably derived from racemic and 2(S)-2-hydroxy-2-(3-chlorophenyl)ethylamine respectively; and (13) from the secondary amine analogue of the latter, by asymmetric reductive alkylation in each case with different carbonyl components. For compound (12), 4-hydroxybenzyl methyl ketone, after conversion to the methoxycarbonylmethyl ether, could be reacted with the foregoing primary amine and the diastereoisomers separated, while with compound (14) the dicarboxymethylene analogue might be employed. For the the tertiary amino derivative (13) the aldehyde 4-(2-ethoxyethoxy)phenylacetaldehyde is a potential component. A variety of synthetic procedures are available for (15) involving the use of (S)-glycidyl phenyl ether and an appropriate 1,4-dioxybenzene derivative.

Compounds for the Control of Appetite

Fenfluramine and Dexfenfluramine [(S)-(+)-Fenfluramine](16) A variety of methods are available for the racemic compound fenfluramine. Thus 3-trifluoromethylbenzaldehyde is convertible to 3-trifluoromethylphenylacetic acid (by several different methods) and thence to the corresponding methyl ketone (this again is available by several alternative methods) which was reductively aminated with ethylamine and the racemic amine (A) resolved[32] with (+)-dibenzoyltartaric acid to give the (−)-salt, leading to (R)-(−)-dexfenfluramine. From the filtrate by further purification with (+)-camphoric acid the enantiomer (S)-(+)-dexfenfluramine was isolated (Scheme 10). The (−)-amine was also obtained from the oxime of the ketone by catalytic reduction, resolution of the primary amine in the same way, and acetylation and reduction of the (−)-enantiomer with lithium aluminium hydride.

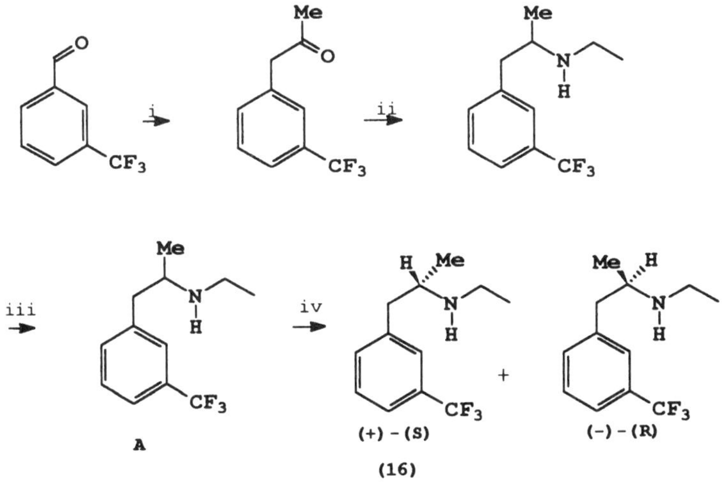

Reactants: (i) $NaBH_4$; $SOCl_2$; ^-CN; HO^-; MeLi, Li salt; (ii) $EtNH_2$, Pd-C, H_2; (iii) (+)-di-PhCO-tartaric acid.

Scheme 10 *Synthesis of dexfenfluramine (16)*

Fluoxetine (17). One route[34] to this compound consisted of the reduction of the Mannich base from the reaction of acetophenone with formaldehyde and dimethylamine with diborane to give the secondary alcohol and thence the chloro derivative was obtained. Reaction with 4-trifluoromethylphenol under basic conditions afforded the ether depicted and conversion to the required *N*-monomethyl product was effected with cyanogen bromide (Scheme 11).

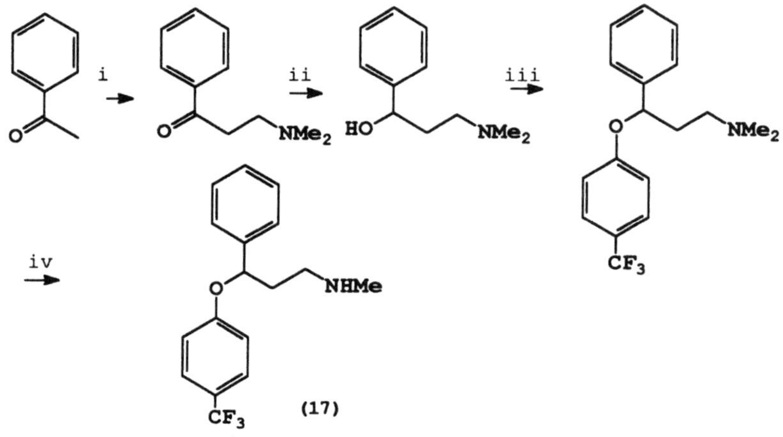

Reactants: (i) B_2H_6; (ii) $SOCl_2$; (iii) 4-$CF_3C_6H_4ONa$; (iv) CNBr.

Scheme 11 *Synthesis of Fluoxetine (17)*

Sertraline (18). One synthesis of this compound proceeds[35] from 3,4-dichlorobenzophenone by Stobbe condensation with diethyl succinate in *tert*-butanol containing potassium *tert*-butoxide. The resultant half ester was hydrolysed and decarboxylated in acetic acid solution with hydrogen bromide and the unsaturated product catalytically hydrogenated, converted to the acid chloride and cyclised to 4-(3,4-dichlorophenyl)tetralone (Scheme 12). The methylimine (Schiff's base) formed with methylamine in toluene solution containing titanium tetrachloride was reduced with sodium borohydride to give a mixture of *cis* and *trans* amines which was separated chromatographically. The *cis* racemate was resolved with (−)-mandelic acid to afford, after basification, the (+)-(1*S*,4*S*) enantiomer, and with (+)-mandelic acid the (−)-(1*R*,4*R*) enantiomer resulted.

Sibutramine (19). In the preparation[37] of this compound, 4-chlorobenzyl nitrile available from the 4-chlorobenzyl chloride was alkylated with 1,3-dibromopropane in the presence of excess sodium hydride to give by cyclisa-

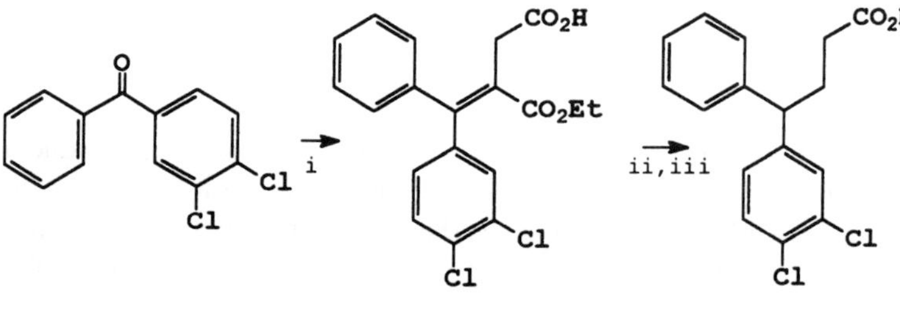

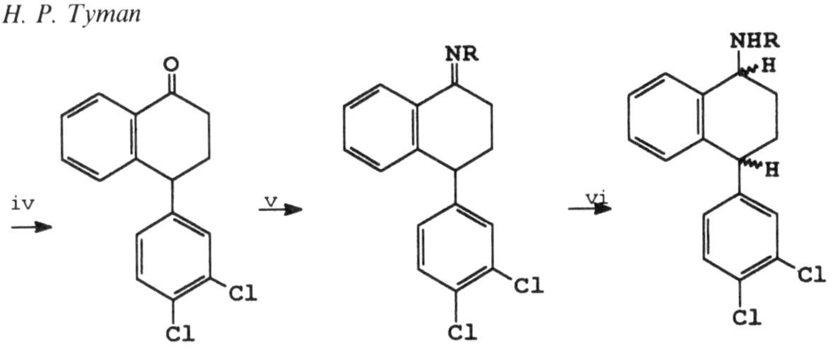

Reactants: (i) $EtO_2C(CH_2)_2CO_2Et$, KO^tBu, HO^-; (ii) AcOH, HBr; (iii) Pd–C, H_2; (iv) $SOCl_2$, $AlCl_3$; (v) $MeNH_2$, MePh, $TiCl_4$; (vi) $NaBH_4$; (−)-$PhCH(OH)CO_2H$.

Scheme 12 *Synthesis of sertraline (18)*

tion the cyclobutane derivative shown (Scheme 13). Grignard reaction with isobutylmagnesium bromide afforded the ketone which with dimethylformamide by heating at 200 °C, afforded the required dimethylamino compound.

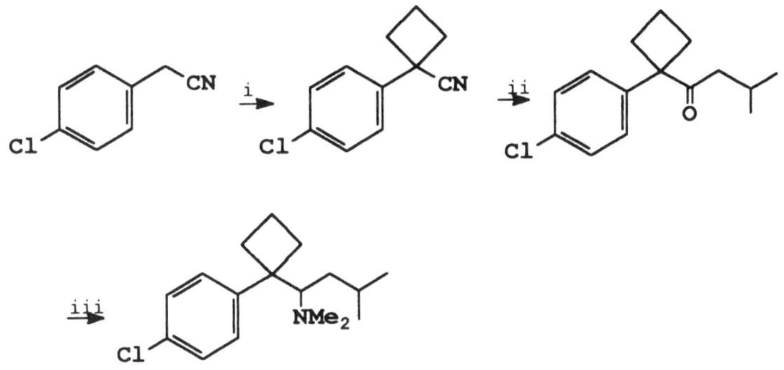

(19) **Sibutramine**

Reactants: (i) $Br(CH_2)_3Br$, NaH, DMSO; (ii) iBuMgBr; H_3O^+; (iii) $HCONMe_2$, Δ, 200 °C.

Scheme 13 *Synthesis of sibutramine*

3-(1H-Indazol-3-ylmethyl)-1,5-benzodiazepine (21). The compound (21) was one of a family of related structures which were synthesised.[38] *N*-Phenyl-1,2-phenylenediamine, obtained readily from 2-nitrodiphenylamine, was reacted with 2-bromo-*N*-isopropyl-*N*-(4-methoxyphenyl)acetamide and the resultant derivative converted to the diazepine with malonyl chloride. The carbanion of this was alkylated with 3-bromomethylindazole to afford the final product (Scheme 14).

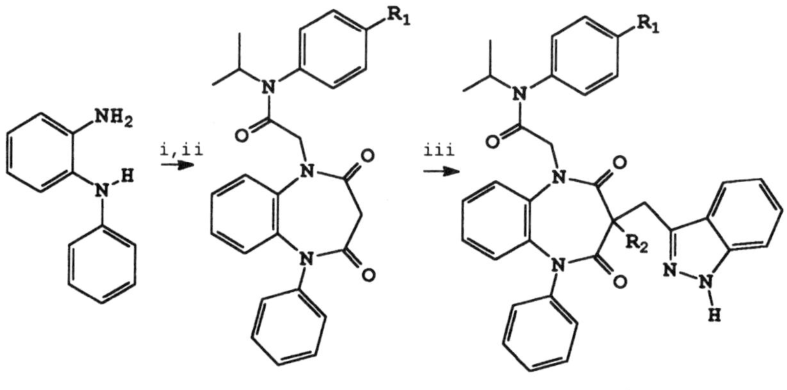

Reactants: (i) BrCH$_2$CON(iPr)C$_6$H$_4$OMe, K$_2$CO$_3$, DMF, 18 h; (ii) CH$_2$(COCl)$_2$, THF, 0 °C to ambient, 18 h; (iii) NaN(TMS)$_2$, ambient, DMF; 3-BrCH$_2$-indazole, 2–5 h.

Scheme 14 *Synthesis of (21)* (R$_1$ = OMe, R$_2$ = H)

Inhibition of Fat Biosynthesis

The resorcinolic compound (22) has been synthesised by a variety of methods[62,66,71] and the isopentadecyl analogue by a related methodology.[72] 3,5-Dimethoxybenzaldehyde reacted with 1-bromotetradecane in tetrahydrofuran in the presence of lithium afforded the corresponding 2 secondary alcohol which was hydrogenolysed with hydrogen in ethanol containing Pd–C to give 1,3-dimethoxy-5-pentadecylbenzene. Demethylation with boron tribromide gave 1,3-dihydroxy-5-pentadecylbenzene. (Scheme 15)

The anacardic acids, compounds (24), (25), (26) and (27), have been synthesised by well-established routes[63] and more recent methodology.[73,74]

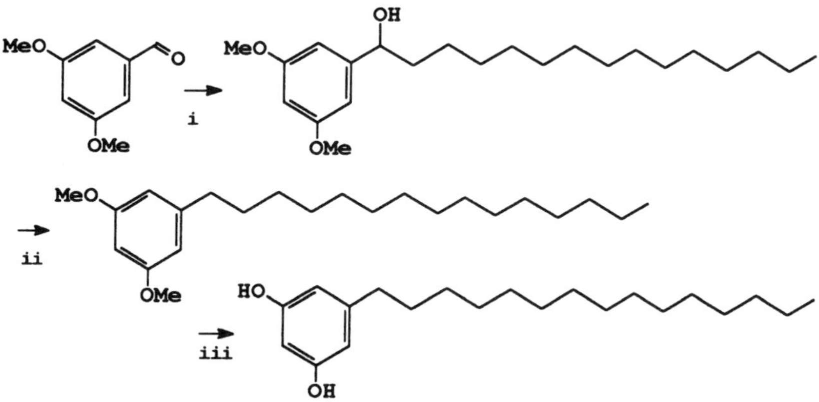

Reactants: (i) C$_{14}$H$_{29}$Br, Li, THF; (ii) Pd–C, H$_2$, H$_3$O$^+$; (iii) HBr.

Scheme 15 *Synthesis of (22)*

References

1. 'Health Survey for England 1994', HMSO, London; M Burrows, *New York Times*, July 17th 1994.
2. M.W. Gillman, L.A. Cupples, B.E. Millen, R.C. Ellison and P.A. Wolf, *J. Am. Med. Assoc.*, 1997, **278**, 2145.
3. T. Sanders, 'Have we Overestimated the Effect of Dietary Fat on Cardiovascular Disease', Lecture to the Oils and Fats Group, Society of Chemical Industry, June 11th 1998.
4. A.M. Prentice and S.A. Jebb, *Br. Med. J.*, 1995, **311**, 437.
5. NIH Consensus Conf. *J. Am. Med. Assoc.*, 1985, **253**, 2080.
6. S.M. Grundy, *West J. Med.*, 1978, **128**, 13.
7. A. Endo, M. Kuroda and Y. Tsujita, *J. Antibiot.*, 1976, **29**, 1346; A. Endo, *J. Med. Chem.*, 1985, **28**, 402.
8. A.G. Brown, T.C. Smale, T.J. King, R. Hasenkamp and R.H. Thompson, *J. Chem. Soc. Perkin Trans 1*, 1976, 1165.
9. N-Y. Wang, C-T. Hsu and C.J. Sih, *J. Am. Chem. Soc.*, 1981, **103**, 6538.
10. A.W. Alberts, J. Chen, G. Kuron, V. Hunt, J. Huff, C. Hofmann, J. Rothrock, M. Lopez, H. Joshua, E. Harris, A. Patchett, R. Monaghan, S. Currie, E. Stapley, G. Albers-Schonberg, O. Hensens, J. Hirshfield, K. Hoogsteen, J. Liesch and J Springer, *Proc. Natl. Acad. Sci. USA*, 1980, **77**, 3957.
11. Nelson and Van den Berg, US 3,692,895, 1969; M. Ast and W.H. Fishman, *J. Chem. Pharmacol.*, 1990, 30, 99.
12. H.R. Casdorph, in *Lipid Pharmacology*, eds. R. Paoletti and C.J. Glueck, Academic Press, New York, 1976, vol. 2, p. 222–256,.
13. P.T. Kuo, K. Hayase, J.B. Kostis and A.B. Moreyra, *Circulation*, 1979, **59**, 199.
14. A.Mieville, Ger. Offen, 2,250,327, 26 Apr. 1973; *Chem. Abs.*, 1973, **79**, 53041.
15. P. L. Creger, Ger. Offen, 1,925,423, 27, Nov. 1969; *Chem. Abs.*, 1970, **72**, 43168.
16. H. Gilman and G.R. Wilder, *J. Am. Chem. Soc.*, 1955, **77**, 6644.
17. E.C. Witte, K. Stach, M. Thiel, F. Scmidt and H. Stork, Ger. Offen, 2,149,070, 5 Apr 1973; *Chem. Abs.*, 1973, **79**, 18438.
18. J.R. Kaplan, *Eur. Heart J.*, 1987, **8**, 928
19. L.J. Theodore and W.L. Wilson, *J. Org. Chem.*, 1987, **52**, 1309.
20. F. Bossert and W. Vater, S. African 68 01,482, 7 Aug. 1968; *Chem. Abs.*, 1969, **70**, 96641
21. R. Paoletti, F. Bernini, R. Fumagalli, M. Allorio and A. Corsini, *Ann. N.Y. Acad. Sci.*, 1988, **522**, 390.
22. M.B. Neuworth, R.J. Laufer, J.W. Barnhart, J.A. Sefranka and D.D. McIntosh, *J. Med. Chem.*, 1970, **13**, 722; R.J. Laufer, US 3,479,407, 18 Nov. 1969; *Chem. Abs.* 1970, **72**, 31445.
23. P. Barbier and F. Schneider, *Helv. Chim. Acta*, 1987, **70**, 196.
24. S. Hanessian, A. Tehim and P. Chen, *J. Org. Chem.*, 1993, **58**, 7768.
25. (a) M.L. Drent, I. Larsson, T. William-Olsson, F. Quaade, F. Czubayko, K. von Bergmann, W. Strobel, L. Sjöström E.A. van der Veen, *Int. J. Obesity Relat. Met. Disord.*, 1995,**19**, 221; M.L. Drent and E.A. van der Veen, *Obesity Res.*, 1995, **3**, 623S; M.L. Drent, C. Popp-Snijders, H.J. Ader, J.B.M.J. Jansen and E.A. van der Veen, *Obesity Res.*, 1995, **3**, 573; (b) R. Guerciolini, *Int. J. Obesity*, 1997, **21**, (suppt. 3), s12.

26. L.F. van Gaal, J.I. Broom, G. Enzi and H. Toplak, *Eur. J. Clin. Pharmacal.*, 1998, **54**, 125.
27. N.J. Rothwell and M.J. Stock, *Nature*, 1979, **281**, 31; D.G. Nicholls, *Physiol. Rev.*, 1984, **64**, 1.
28. J.R.S. Arch, *Proc. Nutr. Soc.*, 1989, **48**, 215.
29. L.J. Beeley, M. Thompson, D.K. Dean, N.R. Kotecha, J.M. Burge and R.W. Ward, PCT Int. Appl. WO 96 04,234, 15 Feb. 1996; *Chem. Abs.*, 1996, **125**, 33308.
30. J.D. Bloom, T.H. Claus, V.G. DeVries, J.O. Dolan and M.D. Dutia, US 5,061,727, 29 Oct, 1991; *Chem. Abs.*, 1992, **116**, 106271.
31. L. Alig and M. Mueller, Eur. Pat. Appl. 198,412, 16 Apr. 1985; *Chem. Abs.*, 1989, **110**, 231262.
32. B.R. Holloway, R. Howe and B.S. Rao, US 5,502,078, 26 Mar, 1996; *Chem. Abs.*, 1996, **125**, 1405x.
33. N. Sasaki, E. Uchida, M. Niiyama, T. Yoshida and M. Sato, *J. Vet. Med. Sci.*, 1998, **60**, 459, 465.
34. E. Savontaus, U. Pesonen, J. Rouru, R. Huupponen and M. Koulu, *Eur. J. Pharmacol.*, 1998, **347**, 265.
35. S.M. Kelly, M. Shorthouse, J.C. Cotterell, A.M. Riordan, A.J. Lee, D.L. Thurnham, R. Hanka and J.O. Hunter, *Br. J. Nutr.*, 1998, **80**, 41.
36. L. Beregi, P. Hugon, J.C. Le Douarec and H. Schmitt, US 3,198,834, 3 Aug. 1965; *Chem. Abs.*, 1965, **63**, 11426.
37. N.E. Rowland and A.J. Don, *J. Physiol. Behav.*, 1995, **58**, 794.
38. *Chem. Ind. (London)*, 1997, 542; 1998, 672.
39. K.K. Schmiegel and B.B. Molloy, US 4,314,081, 10 Jan, 1974; *Chem. Abs.*, 1974, **83**, 192809.
40. W.M. Welch, A.R. Kraska, R. Sarges and B.K. Koe, *J. Med. Chem.*, 1984, **27**, 1508.
41. M.E. Sayler, D.J. Goldstein, P.J. Robck and R.L. Atkinson, *Int. J. Obesity Relat. Met. Disord.*, 1994, **18**, 742.
42. J.E. Jeffery, US 4,929,629; *Chem. Abs.*, 1987, **107**, 223314. .
43. T.M. Willson, B.R. Henke, T.M. Momtahen, P.L. Myers, E.E. Sugg, R.J. Unwalla, D.K. Croom, R.W. Dougherty, M.K. Grizzle, M.F. Johnson, K.L. Queen, T.J. Rimele, J.D. Yingling and M.K. James, *J. Med. Chem.*, 1996, **39**, 2655, 3030.
44. A. Picarelli, Eur Pat. Appn., 0724840; *Chem. Abs.*, 1974, **83**, 192809.
45. H.M. Frankish, S. Dryden, D. Hopkins, Q. Wang and G. Williams, *Peptides*, **16**, 757.
46. C. Gerald, M.W. Walkwe, L. Criscione, E.L. Gustafson, C. Batzi-Hartmann, K.E. Smith, P. Vaysse, M.M. Durkin, T.M. Laz, D.L Linemeyer, A.O. Scaffhauser, S. Whitebread, K.G. Hofbauer, R.I. Taber, T.A. Branchek and R.L Weinshank, *Nature*, 1996, **382**, 168.
47. Y. Zhang, R. Proenca, M. Maffel, M. Barone, L Leopold and J.M. Friedmann, *Nature*, 1994, **372**, 425.
48. J.J. Halaas, K.S. Gajiwala, M. Maffei, S.L. Cohen, B.T. Chait, D. Rabinowitz, R.L. Lallone, S.K. Burley and J.M. Friedmann, *Science*, 1995, **269**, 543.
49. N. Levin, C. Nelson, A. Gurney, R. Vandlen and F. de Savage, *Proc. Natl. Acad. Sci. USA*, 1996, **93**, 1726.
50. T.V. Considine, M.K. Sinha, M.L. Heiman, A. Kriauciunas, T.W. Stephens,

M.R. Nyce, J.P. Ohanessian, C.C. Marco, L.J. McKee, T.L. Bauer and J.F. Caro, *New Engl. J. Med.*, 1996, **334**, 292.
51. J.F. Caro, J.W. Kolaczynski, M.R. Nyce, J.P. Ohannesian, I. Opentanova, W.H. Goldman, R.B. Lynn, P. Zhang, M.K. Sinha and R.V. Considine, *Lancet*, 1996, **348**, 159.
52. M.W. Schwartz, E. Peskind, M. Raskind, M. Boyko and D. Porte, *Nat. Med.*, 1996, **2**, 589.
53. K.L Houseknecht, C.A. Baile, R.L. Matteri and M.E. Spurlock, *J. Animal Sci.*, 1998, **76**, 1405.
54. T.W. Stephens, M. Basinski, P.K. Bristow, J.M. Bue-Valleskey, S.G. Burgett, L. Craft, J. Hale, J. Hoffmann, H.M. Hsiung, A. Kriauclunas, W. MacKellar, P.R. Rosteck, B. Schoner, D. Smith, F.C. Tinsley, X. Zhang and M. Helman, *Nature*, 1996, **377**, 530.
55. F.J. Smith, L.A. Campfield, J.A. Moschera, P.S. Ballon and P. Burn, *Nature*, 1996, **382**, 307.
56. P.S. Widdowson, *Chem. Ind. (London)*, 1997, 55.
57. T. Sakurai, A. Amerniya, M. Jochii, I. Matsuzaki, R.M, Chemelli, H. Tanaka, S. Clay Williams, J.A. Richardson, G.P. Kozlowski, S. Wilson, J.R.S. Arch, R.E. Buckingham, A.C. Haynes, S.A. Carr, R.S. Annan, D.E. McNulty, Wu-Schyong Liu, J.A. Terrett, N.A. Elsghourbagy, D.J. Bergsma and M. Yanagisawa, *Cell*, 1998, **92**, 573; J.S. Flier and E. Maratos-Flier, *Cell*, 1998, **92**, 437.
58. L. de Lecea, T. S. Kilduff, C. Peyron, X.-B. Gao, P.E. Foye, P.E. Danielson, C. Fukuhara, E.L.F. Battenberg, V.T. Gautvik, F.S. Bartlett II, W.N. Frankel, A.N. van den Pol, F.E. Bloom, K.M. Gautvik, and J,G, Sutcliffe, *Proc. Natl. Acad. Sci. USA*, 1998, **95**, 322.
59. L.S. Wise and H. Green, *J. Biol. Chem.*, 1979, **254**, 273.
60. W. Nader, A. Betz and J.U. Becker, *Biochim. Biophys. Acta*, 1979, **571**, 177.
61. G. Klock and K. Kreuzberg, *Biochim. Biophys. Acta*, 1989, **991**, 347.
62. D.J. Mcloughlin, I.I. Shahied and R. Macquarrie, *Biochim. Biophys Acta*, 1978, **527**, 193.
63. (a) J.H.P. Tyman, *Chem. Soc. Rev.*, 1979, **8**, 499; (b) *Synthetic and Natural Phenols*, Elsevier, New York, 1996; (c) *The Chemistry of Non-isoprenoid phenolic Lipids*, in *Studies in Natural Products Chemistry*, ed. Atta-ur-Rahman, Elsevier, New York, 1991, vol. 9.
64. N. Tsuge, M. Mizokami, S. Imai, A. Shimazu and H. Seto, *J. Antibiot.*, 1992, **45**, 886.
65. J. Irie, M. Murata and S. Homma, *Biosci., Biochim.* 1996, **60**, 240.
66. A. Kozubek and J.H.P. Tyman, *Chem. Phys. Lipids*, 1995, **78**, 29.
67. J. Bobowska, J. Sosnowska and Z. Eckstein, *Przem. Chem.*, 1980, **59**, 495; *Chem. Abs.*, 1981, **94**, 102975.
68. P. Barbier, F. Schneider and U. Widmer, *Helv. Chim. Acta*, 1987, **70**, 1412; P. Barbier and F. Schneider, *J. Org. Chem.*, 1988, **53**, 1218.
69. J.-M. Pons and P. Kocienski, *Tetrahedron Lett.*, 1989, **30**, 1833.
70. L.M. Dollinger and A.R. Howell, *Bioorg. Med. Chem. Lett.*, 1998, **8**, 977.
71. J.H.P. Tyman and A. Kozubek, *Chem. Rev.*, 1999, **99**, 1.
72. J.H.P. Tyman and A. Kozubek, unpublished work.
73. J.H.P. Tyman and N. Visani, *J. Chem. Res.*, 1997, 14.
74. J.H.P. Tyman and N. Visani, *Chem Phys. Lipids*, 1997, **85**, 157.

7
Docosahexaenoic Acid: A Dietary Factor Essential for Individuals with Dyslexia, Attention Deficit Disorder and Dyspraxia?

B. J. Stordy

STORDY JONES NUTRITION CONSULTANTS, MANOR HOUSE, PUTTENHAM, GUILDFORD GU3 1AP, UK

1 Introduction

Many factors influence a child's success at school: the home environment, the quality of the teaching at school and the child's personality and response to their surroundings. There is now a substantial amount of evidence that nutrition in utero, in early life and even later in life can influence visual and intellectual development and performance.[1-6] The specific learning disorders dyslexia, dyspraxia and attention deficit/hyperactivity disorder have been linked to altered nutrient requirements, particularly for certain fatty acids.[7-9] In the USA between 1976 and 1993 a threefold increase in prevalence of learning disabilities has been observed,[10] indicating that some environmental factor, possibly dietary, may be contributing.

2 Brain and Retinal Composition

The brain, the most membrane rich tissue in the body, is largely made up of lipids, roughly 60%. Of that lipid, over half is long chain polyunsaturated fatty acids of the $\omega-3$ and $\omega-6$ series, docosahexaenoic acid (DGA) ($\omega-3$) and arachidonic acid (AA) ($\omega-6$) being the most prevalent. Nerve growth cones and synaptic membranes are enriched with DHA, indicating an important role in dendrite arborisation and synaptic transmission.[11,12] DHA and AA are synthesised in the body from the essential fatty acids α-linolenic acid and linoleic acid in food. The retina in man too is made up of cells, rod and cone cells, which have stacks of membrane discs largely composed of phospholipid with long chain polyunsaturated fatty acids (LCPs), particularly DHA as the predominant fatty acid. In the rod cell outer segment there are approximately 1000 discs and 50% of the fatty acid is DHA.[13] The discs are constantly replaced and the complete outer segment is replaced in 10 days to 1 month. The rod cells and some cone cells send messages to the brain through the large neurones of the magnocellular pathway.[14] Interestingly not all species of birds have both rod and cone cells in the retina. The owl has just rod cells. The owl hunts at night and so needs good scotopic vision; it also hunts relatively large prey which move and needs to detect movement over a wide field of vision efficiently. Rod cells and the magnocellular pathway are important for dark adaptation, for movement detection and peripheral vision. On the other hand, pigeons have only cone cells in the retina. They feed on very small seeds and need to distinguish them from stones or other inedible material. The cone cells are important for visual acuity—sharpness of vision and colour vision. Pigeons roost at night because they do not see well in light of low intensity. In dyslexia there is evidence of magnocellular pathway dysfunction.[15,16] Dyslexics have poor movement detection[17] and dark adaptation.[17]

3 Fats in Food and the Metabolism of Essential Fatty Acids

The essential fatty acids, linoleic and α-linolenic acid, are found in plant lipids. Oils derived from seeds and nuts are particularly rich in the $\omega-6$ fatty acid, linoleic acid, and lipids in dark green leafy vegetables and a narrower range of seed oils provide the $\omega-3$ fatty acid α-linolenic acid. In modern diets, roughly half of the total fat content is derived from hydrogenated fats derived from plant oils. The more stable hydrogenated and partly hydrogenated fats have advantages in terms of shelf life but a less desirable consequence is the presence of *trans* fatty acids. These *trans* fatty acids interfere with the metabolism of the essential fatty acids and reduce the availability of long chain polyunsaturated fatty acids for phospholipid membrane synthesis. Many foods such as ice-cream, margarine, snack foods, cakes, cookies and pastry contain hydrogenated fats and so they are difficult to avoid. DHA and AA are synthesised in the body from the essential fatty acids α-linolenic acid and linoleic acid by a series of desaturations and elongations. Although it has been recognised that there is, in general competition between the $\omega-3$ and $\omega-6$ series for the same elongation and desaturation enzymes, there are indications that duplicate sets of enzymes are involved in the mitochondria so that there may be pathological conditions where only one set of enzymes, and thus just one essential fatty acid series, is affected.[18]

The essential fatty acids and their metabolites share many biological roles. Two of the most important are firstly their role as structural and metabolically active components of phospholipid membranes and secondly their role as precursors for prostaglandin and eicosanoid synthesis. Prostaglandins and eicosanoids being physiologically active metabolites influence many biochemical processes.

My interest in this area arose out of my research in infant nutrition and the recognition by others in that field of the importance of breast feeding to visual and intellectual development. The development of premature infants that had been given breast milk from a bottle was followed through to eight years of age and compared with a group of similar infants given formula without LCPs. At age eight those given breast milk had an eight point IQ advantage. Social factors which could have influenced the IQ outcome were accounted for in the analysis of results.[19] More recently a group of researchers from New Zealand has shown that breast fed infants go on to achieve more in school and in later employment.[21] One of the key differences between breast milk and infant formula is that breast milk provides the LCPs, DHA and AA, ready formed, whereas most infant formula only provides the precursor essential fatty acids. Interestingly breast milk is a source of γ-linolenic acid (GLA), an ω−6 fatty acid precursor to AA. Very few other foods provide GLA in significant amounts. The newborn infant has a reduced capacity for elongation and desaturation of the essential fatty acids and so the provision of LCPs at that time of rapid brain growth and visual development is very important. LCPs, particularly DHA, are important constituents for membranes in the rod and cone cells of the retina. One of the interesting challenges in the research has been disentangling the effects of LCP deficiency on vision from those on brain function, as many of the techniques to measure brain function and cognitive development require good vision.

4 The Specific Learning Disorders

There are three closely related learning disorders, dyslexia, attention deficit/hyperactivity disorder (AD/HD) and dyspraxia (developmental co-ordination disorder). The chief characteristic of dyslexia is difficulty with reading, writing and spelling despite normal intelligence and adequate educational opportunity. AD/HD is characterised by inattention, impulsive behaviour and in some cases hyperactivity. Poor movement skills are the chief feature of dyspraxia.[20] However there are many common features to the conditions. Dyslexics frequently share the behaviour traits of AD/HD and also the poor motor skills of dyspraxia.[21,22] The left/right confusion and directional difficulties of dyspraxia are also found in some dyslexics even though many dyslexics are very spatially aware. The conditions do not only share common characteristics, they tend to run in the same families. In large families with this genetic predisposition it is not unusual to find one child with dyslexia, another with AD/HD and yet another with dyspraxia all in the same family. Careful questioning usually reveals a similar pattern in earlier generations.

Dyslexia and Long Chain Polyunsaturated Fatty Acid Supplementation

My research on dyslexia was stimulated by my observation that in a large family with many dyslexics over three generations, those who had been breast fed longest were least affected by their dyslexia. However even prolonged breast feeding, of up to two and a half years, did not prevent the expression of the genetic predisposition completely. I proposed that the LCPs in breast milk are somewhat protective against dyslexia. In order to test the hypothesis I sought a biological measure of reduced amounts of LCPs, particularly DHA, in membranes. I chose dark adaptation, as this function of retinal cone and rod cells might well be dependent on the supply of DHA.

First, we examined dark adaptation in a group of young adult dyslexics, men and women, and compared their scotopic vision with a group of similar but non-dyslexic adults. Dark adaptation was impaired in the dyslexic group, particular in the part of the curve that corresponds to adaptation of rod cells (Figure 1). Then, we compared dark adaptation of a small group of dyslexics

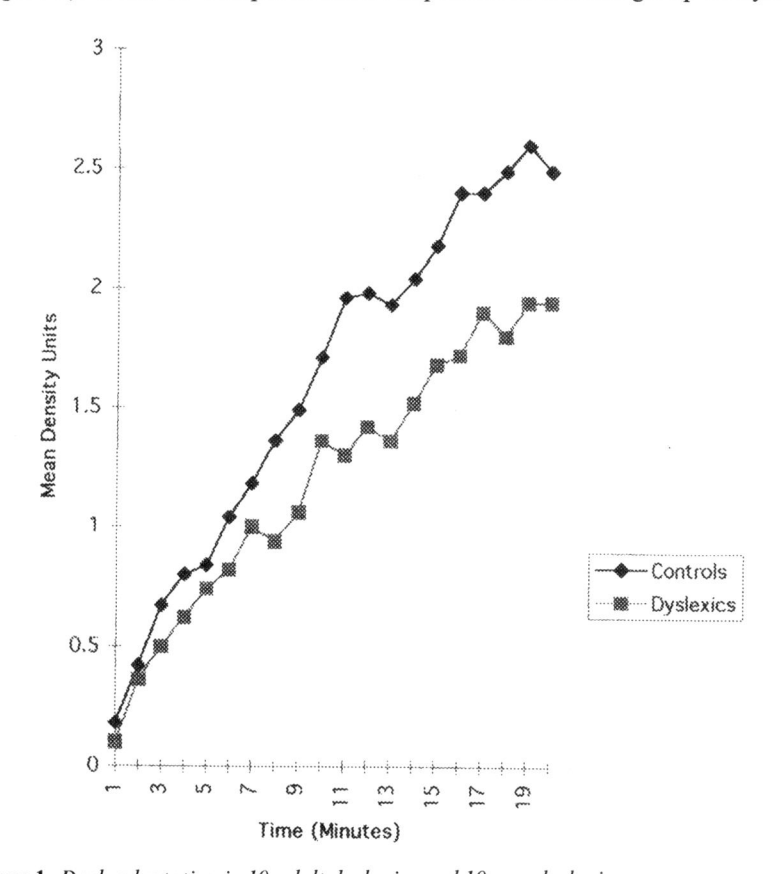

Figure 1 *Dark adaptation in 10 adult dyslexics and 10 non-dyslexics*

with control subjects, in which both groups took a fish oil providing 480 mg of DHA per day for one month. The poor dark adaptation in the dyslexic group improved after one month's supplementation with fish oil but there was no change in dark adaptation in the control, non-dyslexic, group (Figure 2). The improvement in dark adaptation is consistent with the known physiology and biochemistry of phospholipid disc turnover in rod outer segments and suggests that the supply of DHA was limiting the function of rod cells prior to supplementation.

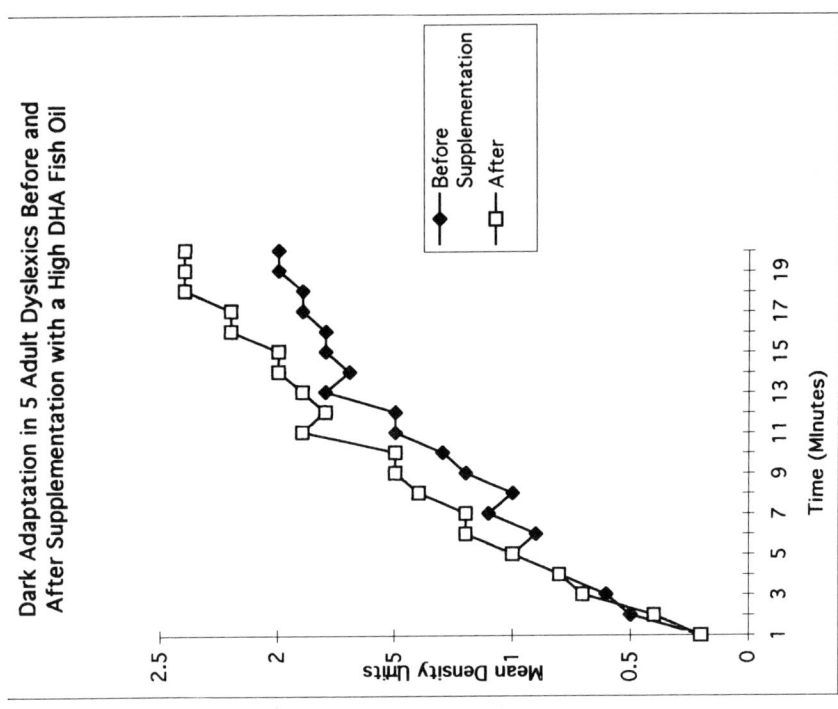

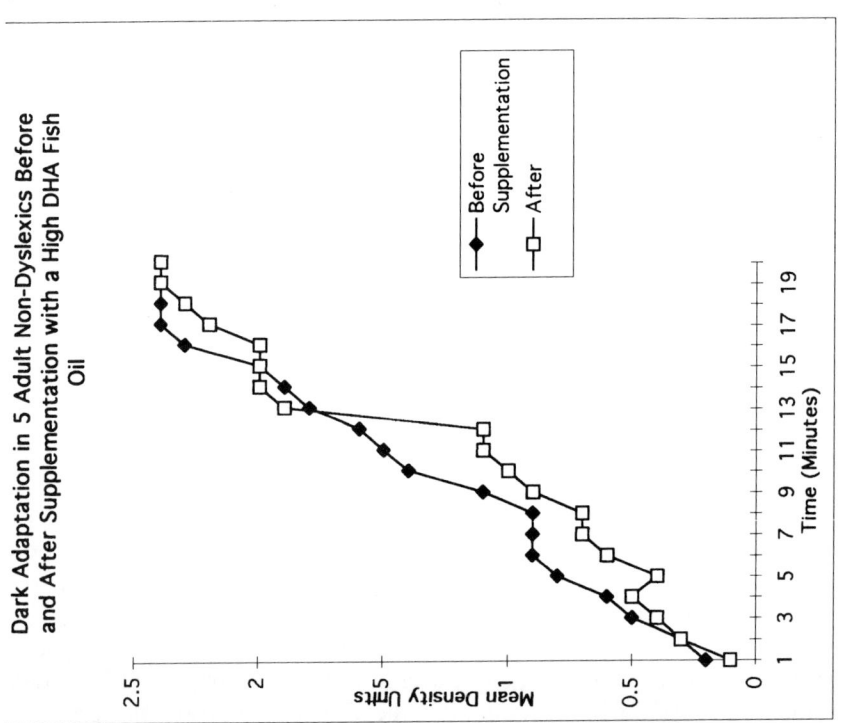

Figure 2

Dark Adaptation in 5 Adult Non-Dyslexics Before and After Supplementation with a High DHA Fish Oil

Dark Adaptation in 5 Adult Dyslexics Before and After Supplementation with a High DHA Fish Oil

Fish oils provide vitamins A and D as well as fatty acids. As vitamin A deficiency is a well known cause of poor dark adaptation,[23] we checked for it as a possible cause of poor dark adaptation in the dyslexic group. Vitamin A intake was estimated from seven day food records completed by both groups. Seven day records are not ideal for this vitamin but it was thought unlikely that dyslexics particularly would complete longer records satisfactorily. There was no difference in vitamin A intakes between the groups and no indication of inadequate supplies of vitamin A from their habitual diet. Interestingly one of the control group was a vegetarian with a low habitual intake of long chain polyunsaturated fatty acids including DHA. Following fish oil supplementation his night vision was much improved. The study was small and has a design which does not allow firm conclusions to be drawn. Larger, double blind placebo controlled, studies are in progress to extend the observations. The baseline data from one of these studies has now been published.[24] The results are consistent with the hypothesis that dyslexics are deficient in long chain polyunsaturated fatty acids.

Fatty acid deficiency signs were investigated in 34 control and 61 dyslexic adults of similar age, sex and general ability and 66 dyslexic children with an age range between 8 and 12 years. In the dyslexic children, more severe reading problems were significantly associated with more severe fatty acid deficiency ($p<0.03$). In adults, signs of fatty acid deficiency were higher in dyslexics than in the control group ($p<0.02$). Specific self report measures of dyslexic symptoms related to auditory-linguistic, motor-co-ordination and visual problems were associated significantly with fatty acid deficiency signs in adults. More severe reading deficits were associated with more severe fatty acid deficiency signs in children (Table 1).

Table 1 *Fatty acid deficiency and reading ability (British ability scales NFER Nelson) in 66 dyslexic children*

	High FAD score mean (SD)	Low FAD score Mean (SD)	Statistical significance	Test used
Reading age (months)	90.4 (10.0)	94.8 (14.2)	ns	Mann-Whitney (data not normally distributed)
Reading lag (months)	30.4 (14.4)	22.7 (13.0)	<0.03	t-test (2-tail)
Reading T-score[a]	37.0 (4.9)	39.8 (5.0)	<0.03	t-test (2-tail)

Data from Richardson *et al.*[24]
[a] Age standardised score with mean = 50, SD = 10. Low scores indicate poor performance.

Brain imaging studies using ^{31}P magnetic resonance spectroscopy have given further support to the hypothesis that there is altered essential fatty acid metabolism in dyslexia.[25] They demonstrated an increase in phospholipid membrane precursors, indicating a shortage of lipids with fatty acids of the preferred composition for brain membrane synthesis.

Attention Deficit/Hyperactivity Disorder and Fatty Acids

In pioneering research[26] on the fatty acid requirements of hyperactive children it was observed that severe thirst was a common feature. While searching through the scientific and medical literature for a biochemical explanation for this the researchers noted that thirst is a feature of essential fatty acid deficiency. This research was continued by Laura Stevens, John Burgess and their group.[4,8] Using a score derived from a four point scale for each of seven items, thirst, frequent urination, dry hair, dandruff, dry skin, follicular keratoses, and brittle nails, they found these clinical signs of essential fatty acid deficiency were more severe in a group of boys with AD/HD (Table 2).

Table 2 *Comparison of possible essential fatty acid deficiency signs in 53 boys with AD/HD and 43 non-AD/HD boys of similar age[8]*

Essential fatty acid deficiency sign	Boys with AD/HD % with symptoms rated in one of the two most severe categories	Control subjects % with symptoms rated in one of the two most severe categories	Statistical significance Fischer's exact two tail test p
Thirst	45.3	16.3	<0.004
Frequent urination	34	7	<0.002
Dry hair	13.2	0	<0.02
Dandruff	7.6	0	ns
Dry skin	11.3	4.7	ns
Follicular keratoses	5.7	11.6	ns
Brittle nails	1.9	2.3	ns
Fatty acid deficiency score >3	39.6	9.3	<0.0009

They also found boys with AD/HD had biochemical evidence of deficiency of the long chain polyunsaturated fatty acids docosahexaenoic acid and arachidonic acid in their red blood cell membranes. The AD/HD group had higher amounts of docosapentaenoic acid of the $\omega-6$ series. This is a well recognised feature of DHA deficiency (Table 3).

They looked for dietary differences between their control group and the AD/HD group which could explain the biochemical differences and found that the AD/HD group actually consumed more of the essential fatty acids, so dietary differences between the control and AD/HD groups could not explain the clinical signs of EFA deficiency. They concluded it may be that the boys with AD/HD are less able to convert the EFAs they consume into the LCPs important for normal biochemical and physiological functions. Interestingly they noted that boys with AD/HD are less likely to have been breast fed and that if they had been breast fed it was for a shorter period of time and the Conners' parent and teacher rating scales were inversely related to the duration of breast feeding (Table 4). This is consistent with our own observations in dyslexia and gives further support to the hypothesis that an adequate provision

Table 3 *Comparison of the fatty acid composition of red blood cell total lipids in boys with AD/HD compared with a control group; values indicate area %*

RBC membrane fatty acids		AD/HD group n = 46	Control group n = 35	p
Saturated				
16:0		15.45	16.44	<0.02
22:0		1.23	0.87	<0.05
Monounsaturated				
18:1		11.26	12.37	<0.05
Polyunsaturated ω−6				
20:4	Arachidonic acid	13.74	15.12	<0.02
22:4	Adrenic acid	4.72	5.21	<0.03
22:5	Docosapentaenoic acid	0.73	0.27	<0.05
Polyunsaturated ω−3				
22:6	Docosahexaenoic acid	1.61	2.18	<0.06

Data from Stevens et al.

Table 4 *Correlation (Pearson correlation coefficients) between behavioural measures, fatty acid composition of red blood cell membrane, fatty acid deficiency score (FADS) and duration of breast feeding in 82–96 young boys*

	FADS	Arachidonic acid	Docosahexaenoic acid	Duration of breast feeding
Conners' scale parent	0.33[a]	−0.14	−0.27[a]	−0.38[a]
Conners' scale teacher	0.23[a]	−0.07	−0.17	−0.27[a]
FADS		−0.37[a]	−0.33[a]	−0.14
Arachidonic acid		1	0.49[a]	0.25[a]
Docosahexaenoic acid			1	0.20[a]

Data from Stevens et al.[8]
[a] $p<0.05$.

of LCPs early in life is somewhat protective against specific learning and behavioural disorders.

Dyspraxia

This condition of many names, including developmental co-ordination disorder, clumsy child syndrome and apraxia, is frequently co-morbid with dyslexia and AD/HD. However, one of the fascinating features of these conditions is that not every individual with dyspraxia has difficulties with reading and writing and not every dyslexic has poor motor skills. Some dyslexics are very talented sportsmen, but some schools teaching dyslexics report that dyslexics make better bowlers than batsmen when playing cricket. Perhaps the visual aspects of dyslexia and poor movement detection contribute to their poor batting skills.

On hearing of the possible value of certain fatty acids for dyslexia, a parent of a dyspraxic child asked me if fatty acid supplements would also help dyspraxics. We initiated a preliminary study to examine this issue. Seventeen parents belonging to a dyspraxia support group volunteered their children for the study. Both children, where possible, and parents gave informed, written consent. The dyspraxic children were between five and twelve years of age and there were four girls and eleven boys in the group. They were tested with the Movement Assessment Battery for Children[27] before and after four months on a fatty acid supplement of tuna oil, evening primrose oil, thyme oil and vitamin E, providing 480 mg DHA, 35 mg AA and 96 mg γ-linolenic acid daily (Efalex, Efamon Lrd, Guildford, UK). It is important to provide an adequate supply of antioxidants with polyunsaturated fatty acids. In the supplement used, vitamin E and various antioxidants in thyme oil such as thymol and carvacrol provided antioxidant protection *in vivo* and *in vitro*.[28] The movement tests used are designed to be used repeatedly to evaluate the benefit, if any, of treatments such as physiotherapy or occupational therapy where it is not possible to hide the nature of the intervention from the participant. There are two main parts to the test, a check list completed by an adult familiar with the child, in this study a parent, and a series of movement skills tests including assessment of ball skills, manual dexterity and static and dynamic balance.

At the outset every child entered the study with a marked degree of movement difficulty; checklist scores for all exceeded the 15th percentile. The movement skills scores were similarly poor. The summed scores for manual dexterity, ball skills and static and dynamic balance form the total impairment score. Population norms are used for comparison. One, 12 year old, child was on the 8th percentile, all the others exceeded the 1st percentile for the total impairment score. After four month's supplementation, statistically significant improvements were seen in the objective measures of movement skills and balance as well as the subjective check list scores (Table 5, Figure 3).

As there is considerable overlap between dyspraxia and AD/HD as well dyslexia, we used Conners' rating scales[29] to evaluate the inattention and behavioural aspects before and after supplementation. There was a statistically significant improvement overall in the Conners' parent scale ($p<0.05$) and a reduction in the anxiety sub-score of the parent scale ($p<0.05$). A trend

Table 5 *ABC Movement Assessment Battery for Children score in 15 dyspraxic children before and after 4 months supplementation with a tuna oil/ evening primrose oil/thyme oil supplement—high scores indicate poor performance*

	Manual dexterity	Ball skills	Static and dynamic balance	Total impairment score	Parent's check list score
Before	9.93	6.03	8.23	24.2	87.14
After	6.94	3.9	5.88	16.73	65.07
p (paired t-test)	<0.01	0.05	<0.05	<0.001	<0.001

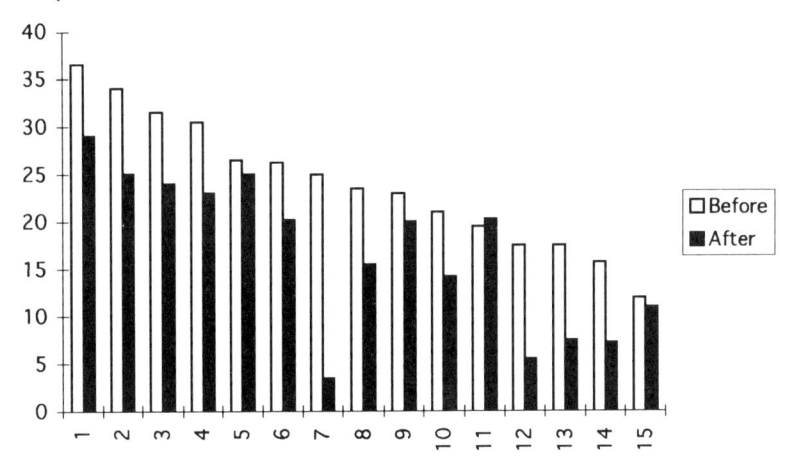

Figure 3 *ABC Movement Assessment Battery Total Impairment Score in 15 dypraxic children before and after supplementation with omega−3 and omega−6 long chain fatty acids*

towards improvement was also seen in the overall Conners' teaching scale ($p<0.1$ but >0.05).

The open design of this study and the complex composition of the supplement does not allow scientific conclusions about the value of individual fatty acids or antioxidants to be made, and much further research is required; but for now it appears reasonable for parents of children with dyspraxia to try fatty acid supplementation. When they do so it is important to use the supplement for at least three months because it takes months, rather than days or weeks, to change brain membrane composition. They should also choose a supplement with a relatively low vitamin A and D content to avoid the dangers of hypervitaminoses. It is useful to have objective tests of movement skills before and after supplement use because of the difficulty of remembering precisely the degree of difficulty a child suffered some months earlier. In practice the movement skills of many such children are regularly monitored by their health care professionals and so it should not be too difficult to implement this proposal.

At the outset of this research I considered docosahexaenoic acid deficiency to be a likely contributor to the visual and central processing deficits in dyslexia. The research of the University of Purdue group[4,8] also suggested the importance of $\omega-3$ fatty acids in learning and behavioural disorders, particularly AD/HD. However, at an NIH workshop ($\omega-3$ Essential Fatty Acids and Psychiatric Disorders, September 1998, Bethesda MD, USA), preliminary results of a placebo-controlled trial in AD/HD with a DHA supplement derived from algae indicated little response, whereas a similar study of supplementation with both $\omega-6$ and $\omega-3$ fatty acids derived from a tuna oil, evening primrose oil and thyme oil supplement produced favourable results. Very preliminary results of a double blind, placebo-controlled trial in dyslexia using the same tuna oil based supplement indicated a trend to improved reading performance and movement skills in a group of severely dyslexic children receiving specialised teaching as well. Thus the indications are that there are benefits associated with fish oil supplements, or fish oil/evening primrose oil/thyme oil supplements, rather than supplements largely providing DHA. There are explanations other than the fatty acid composition of the supplements which could explain the different results of these studies, for example the narrow criteria used to select subjects with AD/HD in the DHA supplementation study and the supply of different amounts of antioxidants, so much more research is needed in this field before sound conclusions can be drawn. Until then the experimental but relatively safe use of supplements based on tuna oil and evening primrose oil may be a worthwhile option for individuals, adults or children, with dyslexia, dyspraxia and attention deficit hyperactivity disorder.

Although it is theoretically possible to manipulate the diet to provide the long chain polyunsaturated fatty acids that appear to be important for dyslexia, AD/HD and dyspraxia, it would be difficult to provide the amounts that appear to be necessary from food. Over 500 g per day of fish would be required. It may be possible to improve the situation somewhat by reducing the consumption of *trans* fatty acids, from hydrogenated and partially hydrogenated fats, and of alcohol, which interfere with the metabolism of essential fatty acids, but so far there is no evidence that this would ensure adequate supplies of the long chain polyunsaturated fatty acids that appear to be conditionally essential for individuals with these specific learning disorders.

Developmental dyslexia has been until recently helped only by multisensory teaching and other educational strategies. Attention deficit/hyperactivity disorder is generally treated with stimulant drugs related to amphetamines together with behaviour therapy and dyspraxia is treated with physiotherapy. However, there appears to be a common biological basis, related to altered essential fatty acid metabolism, for some of the characteristics of the three disorders which responds to dietary intervention with mixtures of long chain fatty acids of the $\omega-3$ and $\omega-6$ series together with antioxidants. The inborn error of fatty acid metabolism together with the environmental dietary change of a higher consumption of hydrogenated and partially hydrogenated oils results in long chain polyunsaturated fatty acids of the $\omega-3$ and possibly the $\omega-6$ series becoming essential nutrients for individuals, children or adults, with specific learning disorders. When conventional and nutritional therapies are used together, these disorders, which cause such difficulty for the individual sufferers and those around them, may be helped.

Acknowledgements

Katrina Searle, Liam Trow and Katy Wood provided invaluable technical skills and active collaboration on the dyslexia dark adaptation studies and Monica Wolff-Kleinman the dyspraxia supplementation study.

References

1. M. Makrides, M. Neumann, K. Simmer, J. Pater and R. Gibson, Are long chain polyunsaturated fatty acids essential nutrients in infancy? *Lancet, 1995*, **345**, 1463–1468.
2. L.J. Horrwood and D.M. Fergusson, Breast feeding and later cognitive and academic outcomes. *Pediatrics 1998*, **101**, 1–7.
3. I. Andraca and R. Uauy, Breast feeding for optimal mental development. *World Rev. Nutr. Diet., 1995*, **78**, 1–27.
4. L.J. Stevens, S.S. Zentall, M.L. Abate, B.A. Watkins, T. Kuczek and J.R. Burgess, Omega-3 fatty acids in boys with behaviour, learning and health problems. *Physiol. Behav., 1996*, **59**, 915–920.
5. R. Uauy, P. Peirano, D. Hoffman, P. Mena, D. Birch and E. Birch, Role of essential fatty acids in the function of the developing nervous system. *Lipids, 1996*, **31**, S167–S176.
6. W.C. Heird, T.C. Prager and R.E. Anderson, Docosahexaenoic acid and the development and function of the infant retina. *Curr. Opin. Lipidol., 1997*, **8**, 12–16.
7. B.J. Stordy, Benefit of docosahexaenoic acid to dark adaptation in dyslexia. *Lancet, 1995*, **346**, 385.
8. L.J. Stevens, S.S. Zentall, J.L. Deck, M.L. Abate, S.R. Lipp and J.R. Burgess, Essential fatty acid metabolism in boys with attention-deficit hyperactivity disorder. *Am. J. Clin. Nutr., 1995*, **62**, 761–768.
9. B.J. Stordy, Dark adaptation, motor skills, docosahexaenoic acid and dyslexia. Paper presented to the Fats of Life Symposium, Barcelona, 1996.
10. W. Roush, Arguing over why Johnny can't read. *Science, 1995*, **267**, 1896–1898.
11. W.C. Breckenridge, G. Gombos and I.G. Morgan, The lipid composition of adult rat brain synaptosomal plasma membranes. *Biochim. Biophys. Acta, 1972*, **266**, 695–707.
12. P.S. Sastry, Lipids of the nervous tissue: composition and metabolism. *Prog. Lipid Res., 1985*, **24**, 169–176.
13. S.J. Fleisler and R.E. Anderson, Chemistry and metabolism of lipids in the vertebrate retina. *Prog. Lipid Res., 1983*, **22**, 79–131.
14. S. Lemkuhle, Neurological basis of visual processes in reading. In: Willows DM, Kruk RS, Corcos E, eds. Visual processes in reading and reading disabilities. Lawrence Erlbaum Associates, Hillsdale, New Jersey, USA. 1993, pp. 77–94.
15. M.S. Livingstone, G.D. Rosen, F.W. Drislane and A.M. Galaburda, Physiological and anatomical evidence for a magnocellular deficit in developmental dyslexia. *Proc. Natl. Acad. Sci. USA, 1991*, **88**, 7934–7947.
16. W. Lovegrove, A. Bowling, D. Badcock and M. Blackwood, Specific reading disability: differences in contrast sensitivity as a function of spatial frequency. *Science, 1980*, **210**, 439–440.
17. G.F. Eden, J.W. VanMeter, J.M. Rumsey, J.M. Maisog, R.P. Woods and T.A. Zeffiro, Abnormal processing of visual motion in dyslexia revealed by functional brain imaging. *Nature, 1996*, **382**, 66–69.
18. J.P. Infante and V.A. Huszagh, On the molecula aetiology of decreased arachidonic (20:4 n-6), docosapentaenoic (22:5 n-6) and docosahexaenoic (22:6 n-3) acids in Zellweger syndrome and other peroxisomal disorders. *Mol. Cell Biochem., 1997*, **168**, 101–115.
19. A. Lucas, R. Morley, T.J. Cole, G. Lister and C. Leeson-Payne, Breast milk and

subsequent intelligence quotient in children born pre-term. *Lancet, 1992,* **339**, 261–264.
20. American Psychiatric Association, Diagnostic and Statistical Manual of Mental Disorders, 4th edition (DSM-IV). Washington DC, 1994.
21. M.N. Haslum, Predictors of dyslexia? *Irish J. Psychol., 1989*, **10**, 622–630.
22. P.H. Wolff, G.F. Michel, M. Ovrut and C. Drake, Rate and timing precision of motor co-ordination in developmental dyslexia. *Dev. Psychol., 1990*, **26**, 349–359.
23. D.S. McClaren, Malnutrition and the eye. Academic Press, New York, 1963.
24. A.J. Richardson, T. Easton, A.C. Corrie, C. Clisby and B.J. Stordy, Is developmental dyslexia a fatty acid deficiency syndrome? Nutrition Society Summer Meeting, Guildford, UK. 30th June–3rd July, 1998.
25. A.J. Richardson, I.J. Cox, J. Sargentoni and B.K. Puri, Abnormal cerebral phospholipid metabolism in dyslexia indicated by phosphorus-31 magnetic resonance spectroscopy. *NMR Biomed., 1997*, **10**, 309–314.
26. I. Colquhoun and S. Bunday, A lack of essential fatty acids as a possible cause of hyerpactivity in children. *Med. Hypotheses, 1981*, **7**, 673–679.
27. S.E. Henderson and D.A. Sugden, Movement assessment battery for children. The Psychological Corporation, Harcourt Brace and Company, London, UK. 1992.
28. R. Aeschbach, J. Loliger and B.C. Scott, Antioxidant actions of thymol, carvacrol 6-gingerol, gingerone and hydroxytyrosol. *Food Chem. Toxicol., 1994*, **32**, 31–36.
29. C.K. Conners, Conners' rating scales manual. Toronto Multi Health Systems, 1990.

8
The Potential for Prostaglandin Pharmaceuticals

D. Clissold

CASCADE BIOCHEM LIMITED, WHITEKNIGHTS PARK, READING, BERKS. RG6 6BX, UK

1 Introduction

Natural prostaglandins (PGs) are biologically potent, short-lived local hormone metabolites derived (Scheme 1), *via* cyclooxygenase (COX) enzymes, from nutritionally essential precursor polyunsaturated fatty acids (PUFAs) and are present in virtually all mammalian tissues.[1] Dietary insufficiency and/or imbalances of precursor PUFAs can cause corresponding disturbances of prostaglandin levels and this is now accepted to be a major factor in a wide range of human ailments and diseases.[1,3]

The PGs are part of a larger family of PUFA metabolites known collectively as the eicosanoids[2] (each molecule contains 20 carbon atoms) which includes the prostacyclins, thromboxanes and lipoxygenase (LOX) derived leukotrienes B_4, C_4, D_4, E_4 and the lipoxins (Scheme 1). Eicosanoids contrast with hormonal products in that they are not stored and do not circulate. Rather they are produced locally on demand, perform a receptor mediated tissue specific function, and then are rapidly inactivated by metabolic enzymes.[3] There have been rapid advances recently in the leukotriene area with the introduction of several antagonists[4] and inhibitors[5] that counter the effects of these pathophysiological mediators in asthma and potentially other allergic conditions.

PGs, of which there are several types (Scheme 2),[1-3] exhibit diverse biological activities controlling a wide range of physiological functions in the circulatory, reproductive, respiratory, digestive and central nervous systems.[1,3] Thus the PGs participate in an extraordinary variety of normal physiological processes[1-3] such as maintaining blood pressure and body temperature, protecting organs from damage caused by disease, traumatic injury and stress, regulating parturition and involvement in sleep–wake cycles. In addition an imbalance in PG levels, brought about by dietary insufficiency of precursor

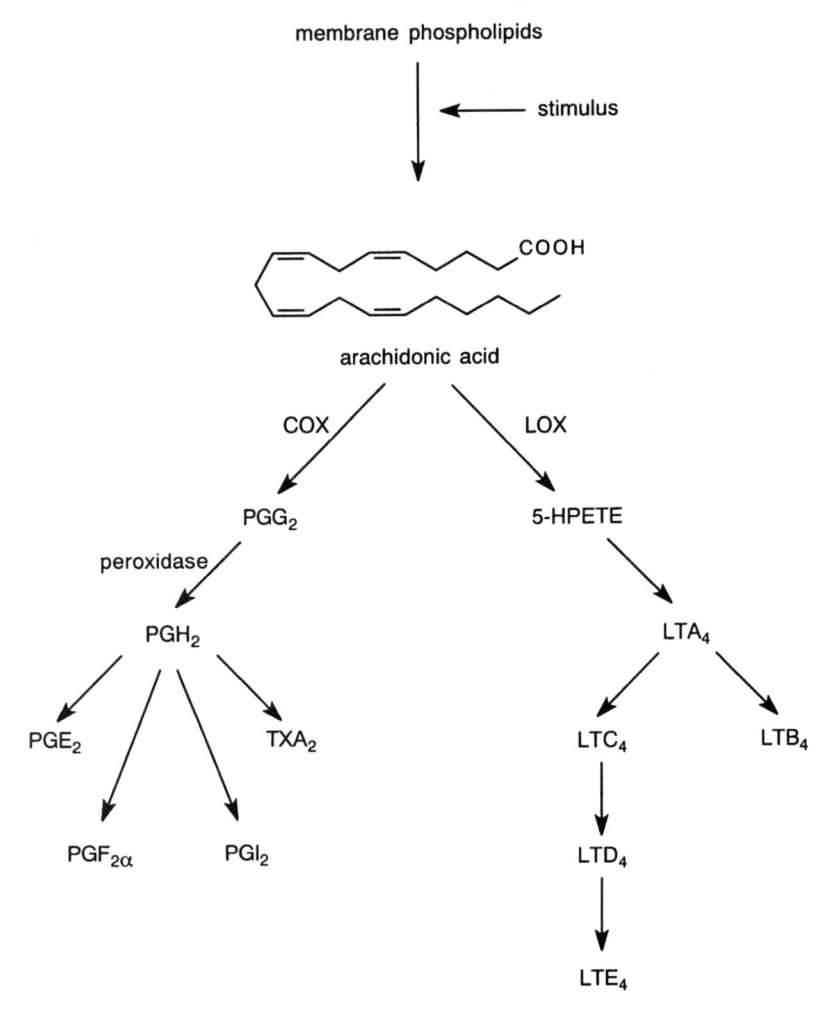

Scheme 1 *The biosynthesis of eicosanoids*
(Note: PGE_1 is produced from dihomo-γ-linolenic acid rather than from arachidonic acid

PUFA or other causes, is implicated in shock and a wide variety of disease states including arthritis, malignancy and allergic disorders.[1-3]

Individual prostaglandins can vary greatly in their activities and potencies; their actions also depend on the animal species, on the tissues in which they are acting and on the concentration present. Entirely opposite actions may be elicited with very small structural changes in the molecule.[1,3]

Given extrinsically, PGs can exert a host of pharmacological effects[1,3] and have therefore been of interest to the pharmaceutical industry for a long period of time[6] and continue so to be.[62]

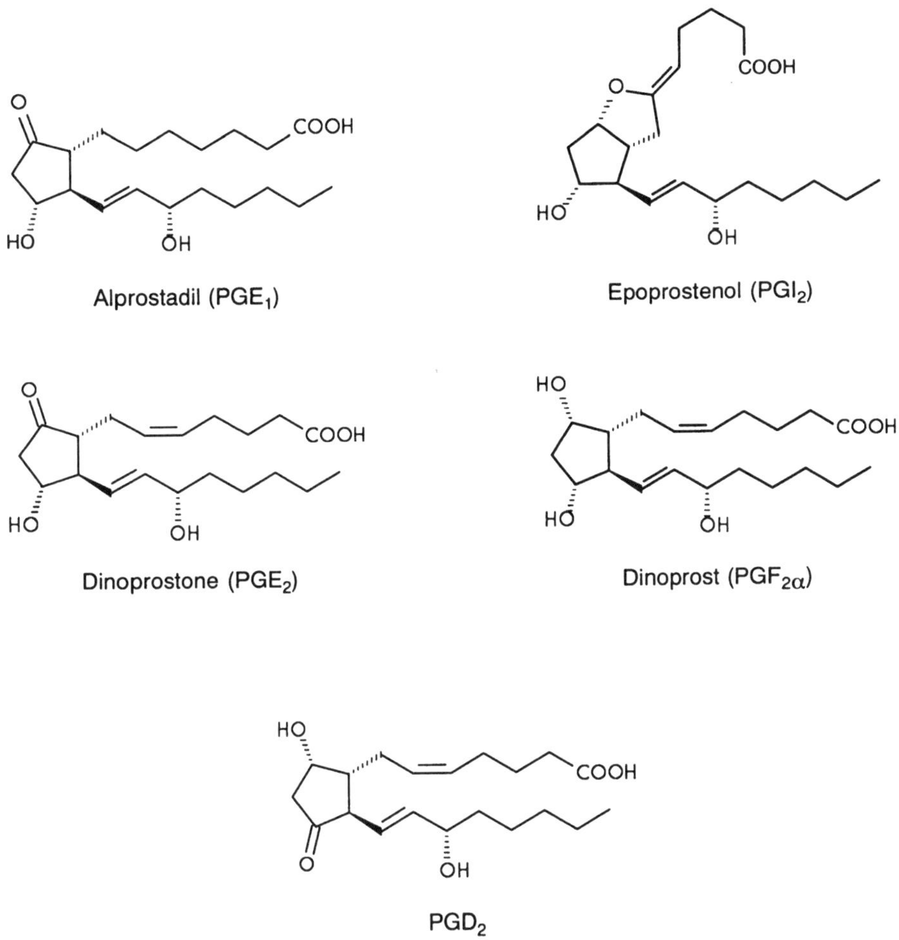

Alprostadil (PGE₁)

Epoprostenol (PGI₂)

Dinoprostone (PGE₂)

Dinoprost (PGF₂α)

PGD₂

Scheme 2 *Important PGs for drug development*

2 Early Drug Development

Although the PGs were discovered[3] in the 1930s by von Euler, isolation and structure determination did not commence until the mid 1950s.[3] The minute quantities of PGs and precursors available from natural sources[6] were inadequate for drug development and consequently synthetic methods were devised, with the first methods appearing more than 30 years ago.[6] With continual refinement over this period, production processes are now available that enable the manufacture of some of these complex molecules in multi kg quantities a year.[7,20] Once reasonable quantities of the synthetic endogenous PGs were available, intensive drug development commenced. However the chemical instability, rapid metabolism and incidence of side effects[6] (Scheme 3) of the natural products impeded the early work and oral delivery was, in general, not feasible.

Chemical instability

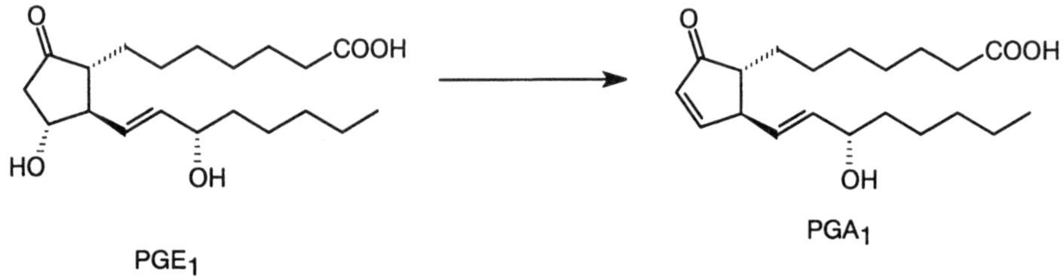

PGE$_1$ → PGA$_1$

- *Rapid metabolism*

β-Oxidation / Reduction / Oxidation

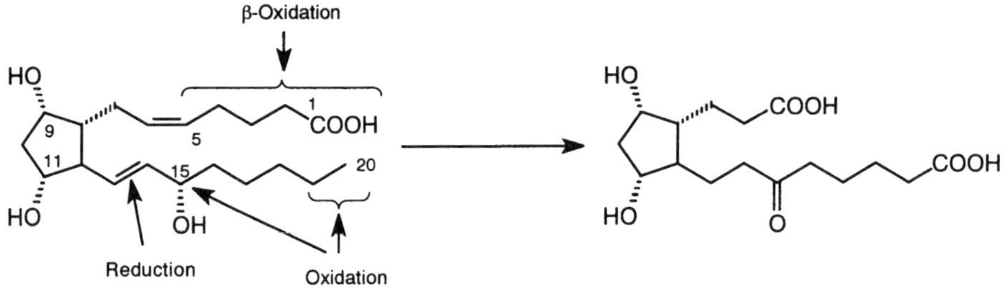

- *Numerous side effects:*
Multiple physiological and pharmacological activities – lack of specificity

Scheme 3 *The problems with prostaglandins*

Although not orally active, natural PGs can be given parenterally and PGE$_1$ (alprostadil), PGE$_2$ (dinoprostone) and PGF$_{2\alpha}$ (dinoprost) and later PGI$_2$ (epoprostenol—as the more stable sodium salt) all became marketed as effective drugs, given by injection or infusion, for a variety of indications (uses) in human and veterinary medicine[6,8] (Scheme 4). Prostaglandin D$_2$ (Scheme 2) has not found utility for drug development but is nevertheless considered to play important roles in sleep–wake cycles[70] and as a mediator in various allergic diseases.[76] PGD$_2$ receptor agonists are being developed for allergic disorders.[76]

Alongside synthetic methodology for making the natural products, techniques were successfully developed[6,7] for making PG analogues usually by adaptation of the methods used for the natural PGs. Although success in finding compounds with enhanced stability, longer duration of action, and with more specific pharmacological actions was initially hard won[6] there were nevertheless, by the mid 1980s, a number of PG analogues on the market. These analogues were launched by several different companies and in some instances were for completely different indications from those for which the parent PG was utilised. A selection of these analogues, with their original indications, is shown in Scheme 5.

Most notably[8] the orally active PGE$_1$ analogues misoprostol[9] and limaprost[10] were introduced for gastrointestinal ulceration and cardiovascular disease respectively; the PGE$_2$ analogues enprostil[11] and arbaprostil[12] were

Prostaglandin	Biological properties	Indications
PGE$_1$ (Alprostadil)	Vasodilator, anti-aggregatory, tissue protection	Congenital heart disease, peripheral vascular disease, erectile dysfunction, female sexual dysfunction, organ transplantation
PGE$_2$ (Dinoprostone)	Smooth muscle contraction, vasodilator, bronchodilator	Induction of labour
PGF$_{2\alpha}$ (Dinoprost)	Vasoconstrictor, bronchoconstrictor, smooth muscle contraction	Fertility control, synchronisation of oestrus (veterinary)
PGI$_2$-N$_\alpha$ (Epoprostenol-Na as sodium salt)	Anti-aggregatory vasodilator	Extracorporeal circulation (e.g. in renal dialysis), primary pulmonary hypertension

Scheme 4 *The biological actions and early applications of natural PGs*

also launched as antiulcer therapies and sulprostone[13] became available for fertility control; the PGF$_{2\alpha}$ analogues carboprost[14] and cloprostenol[15] were marketed for fertility control/post partum haemorrhage and synchronisation of oestrus respectively (the latter for use in veterinary medicine).

The discovery[16] in 1976 of the unstable prostanoid prostacyclin (PGI$_2$) shifted the focus[7] of a number of pharmaceutical companies toward the development of stable analogues of this highly potent vasodilator and anti-platelet aggregatory (anticoagulant) product. Iloprost[17] in Germany and beraprost[18] in Japan were the first stable PGI$_2$ analogue to be marketed. Iloprost is effective for the treatment of ischaemic heart disease and PVD[8] and has been investigated for topical application in skin ulceration.[7] Beraprost[8,18] is another stable prostacyclin which is widely prescribed in Japan for ischaemic conditions such as angina.[8]

3 Changing Markets—Increasing Demand

By the mid-1960s there was widespread belief that the PGs would be useful therapeutic agents in a large number of diseases.[6] However real success was slow to arrive and one or two notable companies ultimately ceased to fund research and development in this area. Thus a perception arose (no doubt due to the long delays in bringing stable selective analogues to market) that the field had become moribund and something of an academic backwater. However in contrast to this there has been a growing recognition of the versatility[19] of the PGs and research and development in the field has continued, with increasing momentum of late. There are now more than thirty PG drugs (including the four natural products alprostadil—PGE$_1$, dinoprostone—PGE$_2$, dinoprost—PGF$_{2\alpha}$ and epoprostenol—PGI$_2$) marketed around

PGE$_1$ analogues	Indications
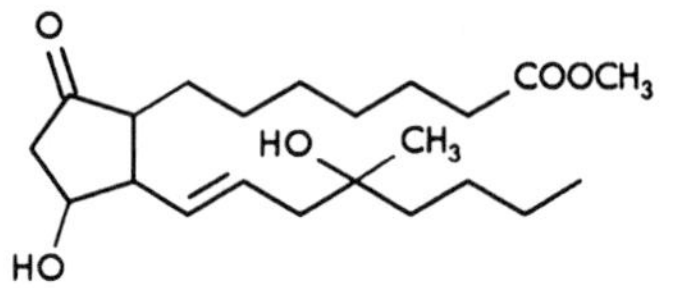 4 stereoisomer mixture misoprostol (oral). SEARLE	Gastrointestinal, ulceration.
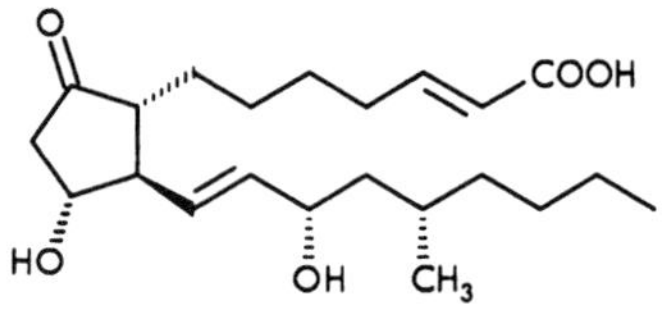 limaprost (oral) ONO	peripheral vascular disease, angina, hypertension.
PGE$_2$ analogues	
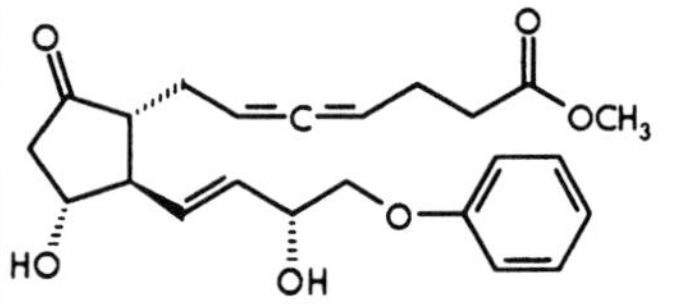 enprostil (oral) SYNTEX	Gastrointestinal, ulceration.
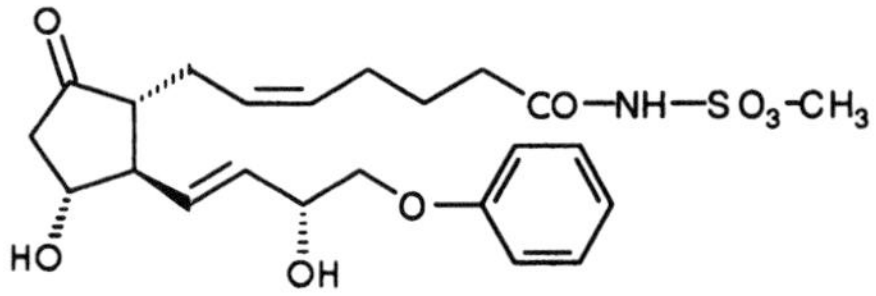 Sulprostone PFIZER/SCHERING	Fertility control.
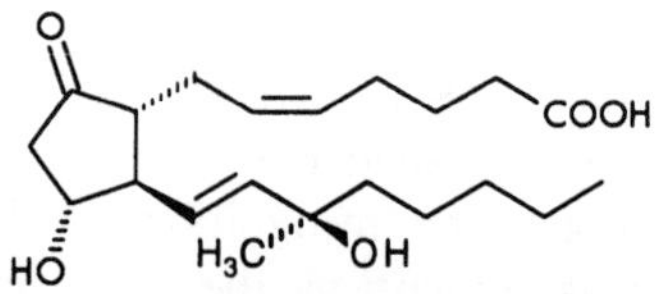 Arbaprostil UPJOHN	Gastrointestinal Ulceration

Scheme 5

PGF$_{2\alpha}$ analogues	
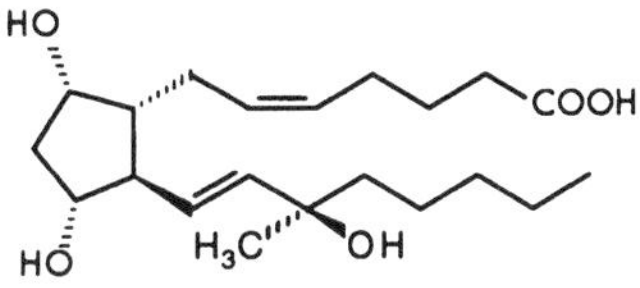 carboprost UPJOHN	Fertility control, primary pulmonary hypertension.
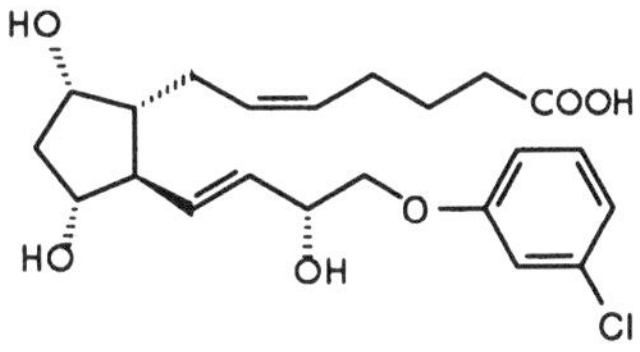 Cloprostenol ICI	Synchronisation of oestrus, luteolytic—veterinary:
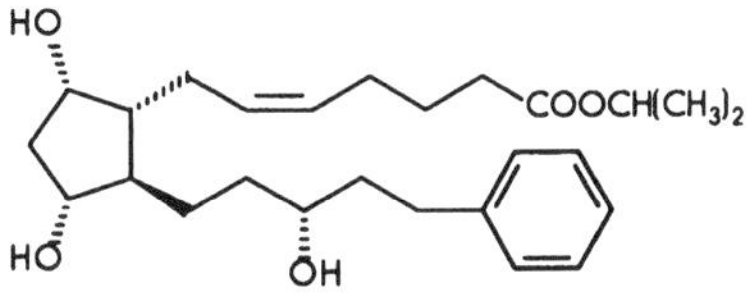 Latanoprost	Glaucoma
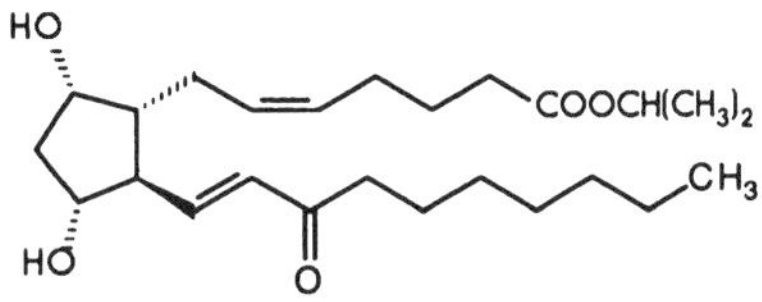 Unoprostone	Glaucoma
PGI$_2$ analogues	
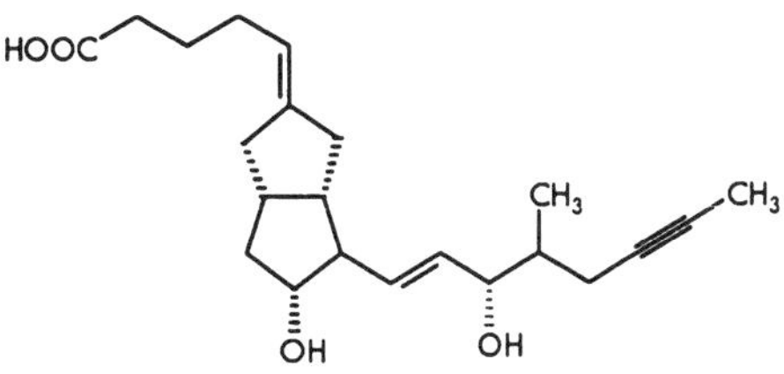 racemic mixture Iloprost (oral, iv, inhalation) SCHERING	Peripheral vascular disease *e.g.* Raynaud's syndrome, peripheral arterial occlusive disease.

Scheme 5 *(continued overleaf)*

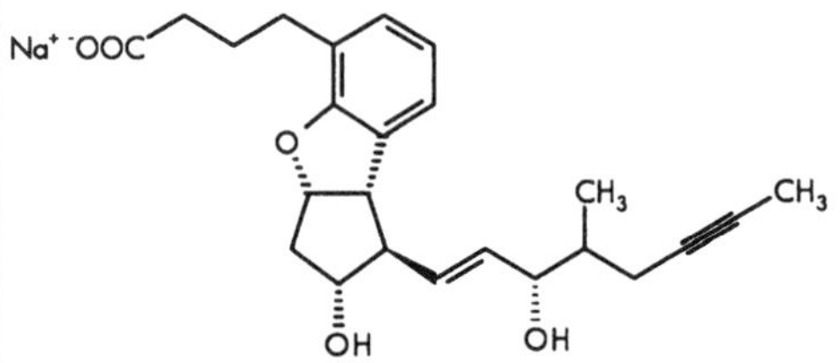 [Structure of Beraprost: Na⁺ ⁻OOC- group attached to benzofuran-cyclopentane with two OH groups, CH=CH, CH₃, and C≡C-CH₃ side chain] 4 stereoisomer mixture **Beraprost**	Cardiovascular disease eg. angina, peripheral vascular disease *e.g.* Raynaud's syndrome, peripheral arterial occlusive disease.

Scheme 5 *Some PG analogue pharmaceuticals*

the world.[8] The top six have a combined and increasing market value, as dosage forms, currently around US$2bn.[20] New stable PG molecules continue to be introduced, sometimes for entirely new uses, and existing PGs are continually being developed for new indications. PG drugs are now established in the key therapeutic areas of cardiovascular, inflammatory and gastrointestinal medicine and are being developed for very large emerging indications such as erectile dysfunction (ED) and female sexual dysfunction (FSD). There also exists a large and growing market for PG products in veterinary medicine. Thus the market has become highly diversified and needs to be viewed from a very wide perspective to appreciate the causes for the growing demand and the overall buoyancy of the field.

One highly significant factor here is the increasing acceptance by the medical profession that the diversity of disorders associated with PG deficiency is wider and of greater pathogenic importance than was earlier recognised.[21] PG imbalances can now be demonstrated in many abnormalities that were earlier described as being of uncertain aetiology.[21] Moreover there is a growing acceptance that total selectivity is not always a prerequisite and that multiple beneficial actions can frequently be utilised to good effect. In tandem with this there is a trend toward the use of combination therapies and novel drug delivery systems and all of these emerging new attitudes, trends and developments have had a positive influence on the PG field. Furthermore the expiry of patent protection on a number of key PG analogues has not missed the attention of the generic and biotechnology industries who are increasingly seeking to add niche, value-added products to their portfolios. These and other influences, discussed later in this article, are summarised in Scheme 6.

4 New PGs, New Indications

More recently, several new PG analogues have been marketed and a number of others are under development. Recent examples include the $PGF_{2\alpha}$ analogues latanoprost[22] and unoprostone[22] (Scheme 5) which have both been launched for the treatment of glaucoma and represent a fundamentally new approach for the treatment of this widespread condition. Of equal importance to the growth of the field is the continuous re-development of existing PGs for

- *Advanced stable, selective analogues*
- *New indications e.g. ED, FSD*
- *Discovery of COX-2*
- *Application of new drug delivery systems*
- *Combination therapies*
- *Patent expiries*

Scheme 6 *Recent positive developments in the PG field*

new indications. Since the PGs have such diverse biological effects it is perhaps not surprising that a given PG drug can be re-invented (potentially many times over) in this way.

A good example of such re-development is the orally stable PGE_1 analogue misoprostol[9] (Scheme 5). This product, which will lose patent protection in 2000, has gained regulatory approval and is licensed in more than 40 countries for the treatment and prevention of NSAID (non-steroidal anti-inflammatory drug) induced gastro-intestinal lesions. However the potential applications of misoprostol go far beyond this original use and draw attention to the unique and extensive scope for further development[23] possessed by this and other PGs.

A recent review[23] 'Novel Applications of Misoprostol' indicates[26] some ten separate uses/indications for which misoprostol could be further developed. These include inflammation, cardiovascular disease, asthma, cystic fibrosis, cancer, septic shock, renal function, liver injury, dermatitis and periodontal disease. An additional area where misoprostol is currently undergoing intense clinical investigation is that of obstetrics[24] and in this arena misoprostol is now emerging as the drug of choice in several indications including induction of labour.[25]

Alprostadil is also being developed for a host of different indications[26] where oral stability is not a requirement and it is conceivable that in any indication where alprostadil is utilised (*e.g.* ED[44] and FSD[47]) misoprostol, as a stable PGE_1 analogue, may also represent an effective therapy.

The oral route for PGE_1 is precluded since it has a very short half-life in vivo.[27] In contrast, limaprost,[10,28,29] another PGE_1 derivative, was designed to be much more chemically stable and resistant to metabolism. As such it is widely used in Japan as an oral drug for the treatment of Raynaud's phenomenon and for ulcers in the extremities caused by chronic occlusive arterial disease.[30-32] Thus the peripheral blood circulation is improved by administration of limaprost and by this mechanism has proved effective *via* the oral route in ED.[33] More recently[34] limaprost was found to be superior to gosyajinki-gan (a widely used Chinese herbal drug) in the treatment of mild

ED. This is an interesting development since in contrast to Viagra, which needs to be prescribed with caution[44] to ED patients with certain heart conditions, limaprost is a cardiovascular drug in its own right and would therefore be expected to be safe in such patients.

Epoprostenol is an example of a PG drug whose second indication has become far more significant than the first. Its original and continuing use is as an adjunct in renal dialysis, charcoal haemoperfusion and cardiopulmonary by-pass.[8] However it has been found to be highly efficacious in the treatment of primary pulmonary hypertension (PPH) and this is now by far the main use for this PG.[35] Similarly the prostacyclin analogues beraprost[36] and iloprost,[37] although developed for other uses,[8] are now also undergoing clinical trials for PPH. In addition iloprost is in clinical trials for Beurger's disease[38] and beraprost is under investigation in organ transplantation.[39]

5 Drug Delivery Systems Applied to PGs

Historically the pharmaceutical industry has focussed on the development of stabilised selective PG analogues to achieve therapeutic goals. However the application of novel drug delivery technologies[40] to PGs is proving to be particularly effective and new products are emerging that effectively overcome the innate problems[3,6] associated with the PGs. A good case in point is PGE_1 (alprostadil) which is directly involved as a natural vasodilatory mediator[41] in the physiological mechanism of erection. This has been taken advantage of in the successful development of intracavernous[42] (Caverject, Edex) fine needle injections and intraurethral[43] (MUSE) delivery systems for the treatment of impotence (erectile dysfunction, ED). These products[44] are marketed and compete alongside the highly successful non-PG oral medicine Viagra.[44] Since the launch of this product by Pfizer, in April 1998, the demand for effective treatments for ED is now recognised to be enormous[44] (Viagra alone will achieve sales in excess of $1bn in its first year). This has spurred on the development of alternative alprostadil based user friendly local delivery systems, including needleless injection[40,45] and topical[46] delivery systems. The latter is a highly active area and there are a number of companies with patented topical delivery systems[44] who will soon bring to market alprostadil based products consisting of transdermal gels, patches and liposome formulations. Alprostadil also promises to be an effective agent[47] in the related area of female sexual dysfunction (FSD) and formulations (mainly topical) are currently being developed[47] by a number of companies for this correspondingly very large (cf. ED) market for which there are currently no drug treatments available.[47]

Small variations in the prostaglandin structure can have dramatic effects on efficiency in transdermal systems and on stability. For example recent work[48] has shown that simple esters such as PGE_1 ethyl ester (a lipophilic pro-drug of PGE_1) penetrates the skin much more efficiently than PGE_1 (a carboxylic acid). The ester is hydrolysed by esterase enzymes[49] on passage through the skin, to give high levels of PGE_1 in the circulation.[48] Such PGE_1 derivatives will no doubt be developed as transdermal treatments for ED and for other indications such as cardiovascular disease where PGE_1 is currently administered intravenously[26] in a hospital settings. The isopropyl ester derivatised PGs latanoprost and unoprostone have already found utility in glaucoma[22] and PGD_2 methyl ester has been found to readily transport into the brain[50] as compared to PGD_2. Similarly viprostol, a methyl ester analogue of dinoprostone, was earlier developed as a topical treatment for PVD and male pattern baldness.[51] Moreover misoprostol (a methyl ester pro-drug[23] for the active free acid metabolite) can be utilised topically in, for example, the treatment of periodontal disease, mouth ulcers[52] and atopic dermatitis.[53]

Stabilty of PGs can also be improved by novel formulation techniques. For example an inclusion complex of a 3% w/w alprostadil in α-cyclodextrin (alprostadil alphadex) has been developed as a stabilised formulation for the treatment of PVD.[52] More recently alprostadil alfadex has found utility for intracavernous treatment of ED[54] and recent reports have indicated that it has potential in needleless injection systems.[40,45]

6 NSAIDs, COX-2 and PGs

The importance of the roles of PGs in inflammatory disease and pain was highlighted by the discovery,[55] in 1971 by Vane, that aspirin and similar (non-steroidal anti-inflammatories, NSAIDs) drugs such as naproxen, ibuprofen, indomethacin and diclofenac produce their anti-inflammatory and analgesic effects by inhibition of cyclooxygenase (COX) enzymes that give rise to pro-inflammatory PGs such as PGE_2. As a side effect all NSAIDs also inhibit the production of tissue protective PGs such as PGE_1 and PGI_2 to a greater or lesser degree. With chronic administration, as is necessary in for example osteo and rheumatoid arthritis, depletion of tissue protective PGs can be very pronounced and particularly severe in the gut and kidney. Thus the price to pay for the beneficial anti-inflammatory and analgesic effects of NSAIDs is the particular risk of inducing ulcerative damage to gastrointestinal mucosa with the subsequent possibility of bleeding, perforation and death. There is thus a rationale for combining an orally stable PGE_1 analogue with NSAIDs to replenish protective prostaglandins depleted by the unwanted component of the NSAID actions.

Misoprostol is such an analogue and has gained regulatory approval in more than 40 countries for the treatment and prevention of NSAID induced gastro-intestinal lesions.[23] Misoprostol is marketed, by Searle-Monsanto, as 'Cytotec' for the above use and as 'Arthrotec' in combination with the NSAID diclofenac for the treatment of inflammatory disease.[23] The latter is a highly effective combination product and enjoys increasingly wide usage around the world as an effective and safe treatment of inflammatory disease. The extent of use of these products can be gauged from the fact that in excess of 150 kg p.a.[20] misoprostol raw material is produced by Searle Monsanto (the inventor of misoprostol) for use in tablets that contain only 100 or 200 μg of the PG component.

Further attention has recently been drawn to the PG field by the discovery[57] that the COX enzyme system exists in two isoforms, COX-1 and COX-2. The constitutive form COX-1 is considered to be responsible for the production of tissue protective PGs while the inducible COX-2 produces pro-inflammatory PGs.[58] Existing NSAIDs are non-selective and inhibit both enzymes, to varying degrees, whereas selective COX-2 inhibitors, currently under development,[59] are anticipated to be ulcer-free treatments. However attention has recently been drawn[60] to several inconsistencies in this elegant theory resulting from experimental work with the new COX-2 inhibitors. In particular the observations[60] that PGs produced from COX-2 may have a beneficial effect on ulcer healing and that COX-1 PGs may contribute toward inflammation are at odds with the theory and have cast some doubt as to how universally useful the new COX-2 inhibitors are likely to be in practice.[60]

With this background, effective PG-NSAID combinations, such as misoprostol/diclofenac,[56] promise to retain a prominent position in the treatment of pain and inflammatory disease in a market for prescription arthritis drugs set to rise[61] to ca. US$19bn by 2005.

Interestingly, misoprostol, like its parent PGE_1, has been shown to have pronounced anti-inflammatory and analgesic properties in its own right[23,56] and exhibits strong synergistic effects with diclofenac.[56] Thus the actions of misoprostol go well beyond simple replenishment of depleted tissue protective PGs[56] and the drug has many beneficial pharmacological actions, in addition to its use as an anti-ulcer medicine, that are currently under pre-clinical and clinical investigation.[23,24]

7 Other Areas of PG Research

The literature on the PGs is vast and growing with currently around 200 papers a month[62] being published. Thus in addition to those topics referred to above there are a growing number of on-going and new areas of research of relevance to PG drug development. In a short review such as this these areas can, unfortunately, only be mentioned in passing. References to review articles are provided wherever possible:

- Roles of PGs in pregnancy[63a] and parturition[63b]
- Involvement of COX enzymes and PGs in cancer of the colon,[63] breast,[64] lung,[65] prostate,[66] skin[67a] and brain.[67b] Selective prevention of cytoxicity of anti-cancer drugs[68]
- Involvement of COX enzymes and PGs in Alzheimer's disease[69]
- Roles of PGs in sleep–wake cycles[70]
- Use of PGs in organ transplantation[71]
- Involvement of isoprostanes as mediators in oxidation injury[72]
- Role of PGs and COX enzymes in the CNS,[73] brain injury[74] and depression[75]
- Roles of PGs in allergic disease[76]

8 Conclusion

Endogenous prostaglandins (PGs) are of fundamental importance in human and animal health and have been a focus of attention for drug development for many years. There are now some thirty PG drugs (including natural PGs and prostacyclin and its analogues) marketed throughout the world with many more under development. PGs are now utilised in the key therapeutic areas of cardiovascular, inflammatory, and gastrointestinal medicine and the top half dozen PG products have a combined market value in excess of $2bn. Current pre-clinical and clinical research will lead, in the foreseeable future, to the introduction of an increasingly diverse range of innovative PG products for use in many areas of medicine.

References

1. J.L. Ninnemann, in *Prostaglandins, Leukotrienes and the Immune Response*. Cambridge University Press, Cambridge, 1988.
2. D. Clissold and C. Thickitt, *Natural Product Reports*, 1994, **11**(6), 621–638.
3. Von Euler, *U.S. Arch. Exp. Pathol. Pharmacol.*, 1934, **175**, 78; J.R. Vane, *Angew. Chem.*, 1983, **95**, 782–794; *Angew. Chem. Int. Ed. Engl.*, 1983, **22**, 741–752; B. Samuelsson, *ibid.*, 1983, **95**, 854–864 and 1983, **22**, 805–815; S. Bergström, *ibid.*, 1983, **95**, 865–873 and 1983, **22**, 858–866; see also: *Adv. Prostaglandin, Thromboxane, Leukotriene Res.*, 1983–1995, **11–23**.
4. B. J. Lipworth, *The Lancet*, 1999, **353**, 57–62.
5. D.M. Shaffer and P.T. Munsmann, *Pediatr. Asthma Allerg. Immun.*, 1997, **11**(4), 171–179.
6. N.A. Nelson, R.C. Kelly, and R.A. Johnson, *Chem. Eng. News*, 1982, **60**, 30–44.
7. R. Noyori, *Chem. Br.*, September 1989, 883–888; P.W. Collins and S.W. Djuric, *Chem. Rev.*, 1993, **93**, 1533–1564.
8. Details on the uses of pharmaceutical prostaglandins can be found in *Martindale, The Extra Pharmacopoeia*, 31st edn., pp. 1449–1462.
9. P. W. Collins, *Med. Res. Rev.*, 1990, **10**, 149–172.
10. S.R. Kottegoda *et al.*, *Prostaglandins Leukotrienes Med.*, 1981, **8**, 343.
11. Series of articles on pharmacology, clinical efficacy and safety: *Am. J. Med.*, 1986, **81**(Suppl. 2A), 1–88.
12. D.A. Gilbert *et al.*, *Gastroenterology*, 1984, **86**, 339.
13. K. Schmidt-Gollwitzer, *Int. J. Fertil.*, 1981, **26**, 86.
14. M.P. Mapa *et al.*, *Int. J. Gynaecol. Obstet.*, 1982, **20**, 125.
15. D. Binder *et al.*, *Prostaglandins*, 1978, **15**,773.
16. J.R. Vane *et al.*, *Nature*, 1976, **263**, 663.
17. S.M. Grant and K.L. Goa, *Drugs*, 1992, **43**, 889–924.
18. R. Kato *et al.*, *Jpn. J. Clin. Pharmacol. Ther.*, 1989, **20**, 529
19. The Versatile Prostaglandin in *Pharma Business*, September/October 1995; J. Thierauch *et al.*, *J. Hypertens.*, 1993, **11**(12), 1315–1318.
20. Data supplied by Kilochem service of International Medical Statistics (IMS), London.
21. M. Moran *et al.*, *Prostaglandins, Leukotrienes Essent. Fatty Acids*, 1990, **39**, 83–89.

22. A. Alm, *Prog. Retina Eye Res.*, 1998, **17**(3), 291–312.
23. M. J. Shield, *Pharmacol. Ther.*, 1995, **65**, 125–147.
24. A. Templeton, *Br. J. Obstet. Gynaecol.*, 1998, **105**(9), 937–939.
25. C.D. Adair and J.W. Weeks, *Obstet. Gynaecol.*, 1998, **92**(5) 810–813.
26. S.J. Kirtland, *Prostaglandins, Leukotrienes Essent. Fatty Acids*, 1988, **27**, 271.
27. M. Gloub, P. Zia , M. Matsuno, R. Horton, *J. Clin. Invest.*, 1975, **56**, 1404–10.
28. T. Tsuboi , B. Fujitani , K. Maeda , *et al.*, *Thromb. Res.*, 1980, **20**, 573–80.
29. H. Ohno, Y. Morikawa, F. Hirata, *J. Biochem.*, 1978, **84**, 1485–90.
30. C. Murai, T. Sasaki, H. Osaki, *et al.*, *The Lancet*, 1989, **2**, 1218.
31. A. Takeuchi, T. Hashimoto, *Int. J. Clin. Pharm. Res.*, 1987, **7**, 283–9.
32. Y. Saito, Y. Kitani, T. Fujita, *Biomed. Thermography*, 1987, **7**, 137–9.
33. Y. Sato, Y. Kumamoto, N. Suzuki, *et al.*, *Impotence*, 1991, **6**, 349–55.
34. Y. Sato, H. Hotita, H. Adachi, *et al.*, *Br. J. Urol.*, 1997, **80**, 772–775.
35. G. Montalescot and G. Drobinski, *Am. J. Cardiol.*, 1998, **82**(6), 749–755.
36. X.W. Jiang, K. Kambara, N. Gotoh, K. Nishigaki, H. Fujiwara, *Am. J. Respir. Crit. Care Med.*, 1998, **158**(5), 1669–1675.
37. T.W. Higenbotham and A.Y. Butt, *Heart*, 1998, **79**(2), 175–179.
38. M. Verstraete, *Eur. J. Vasc. Endovasc. Surg.*, 1998, **15**(4), 300–307.
39. Y. Okada, A.M. Marchevsky, R.M. Kass, J.M. Matloff, S.C. Jordan, *Transplantation*, 1998, **66**(9), 1132–1136.
40. P. Berressem, *Chem. In. Brit*, 1999, **35**(2), 29–32; Sameena Ahmed, "A hi-tech boom helps the medicine go down" in, *The Independent*, 10 Jun 1997; G. Cerc, *Exp. Opin. Invest. Drugs*, 1997, **6**, 1887–1937.
41. H. Porst, *Adv. in Prostaglandin, Thromboxane Leukotriene Res.*, 1995, **23**, 539–544.
42. O.I. Linet and L.L. Neff, *Clin. Investig.*, 1994, **72**, 139–149.
43. H. Padma-Nathan, *et al.*, *Int. J. Impotence Res.*, 1994, **6** (Suppl.1).
44. N. Osterweil in *Beyond Viagra, Drugs, Devices and Markets*. FT Healthcare Management Report, London, 1998.
45. *Urology: an up-and-coming area.* UBS Global Research (German Equities) 1996.
46. E.D. Kim and K.T. McVary, *J. Urol.*, 1995, **153**, 1828–1830.
47. P. Moyer in *Female Sexual Dysfunction, Diverse Types, Causes and Therapies*. FT Healthcare Management Report, London, 1998.
48. M.-T. Sheu, L.-H. Lin, B.W. Spur, P. Y.-K. Wong and H.-S. Chiang, *J. Controlled Release*, 1998, **55**, 153–160.
49. G.L. Bundy, D.C. Peterson, J.C. Cornette, W.L. Miller, C.H. Spilman, J.W. Wilks, *J. Med. Chem.*, 1983, **26**(8), 1089–1099 and references therein.
50. F. Suzuki, H. Havashi, *Biochim. Biophys. Acta*, 1987, **917**(2), 224–230.
51. J.J.F. Belch, *et al.*, *The Lancet*, 1985, **i**, 1180–3; D.B. Dunger, *et al.*, *The Lancet*, 1985, **ii**, 50.
52. T. Olivry, E. Guaguere, and D Heripret, *J. Dermatol. Treatment*, 1997, **8**(4) 243–247.
53. F. Mantovani, F. Colombo, E. Austoni, *Adv. in Prostaglandin, Thromboxane Leukotriene Res.*, 1995, **23**, 545–554.
54. *Martindale, The Extra Pharmacopoeia*, 31st edn., p. 1450.
55. J.R. Vane, *Nature*, 1971, **231**, 232–235.
56. M.J. Shield, *J. Rheumatol.*, 1998, **25**(51), 31–41.
57. W.L. Xie, J.G. Chapman, *et al.*, *Proc. Natl. Acad. Sci. USA*, 1991, **88**, 2692–2696.

58. J.R. Vane, *Nature*, 1994, **367**, 215–216.
59. C.J. Hawkey, *The Lancet*, 1999, **353**, 307–14.
60. J.L. Wallace, *Trends Pharmacol. Sci.*, 1999, **20**, 4–6.
61. *Therapy Markets - Major Areas of Unmet Medical Need to 2005: in Decision Resources Reports*, September 1997.
62. *Current Awareness in Biomedicine Prostaglandins Biology*, Sheffield Academic Press (SUBIS twice monthly prostaglandin literature survey).
63. (a) H. Wunch, *Lancet*, 1998, **351**(9119), 1864; G.R. Davies and D.S. Rampton, *Eur. J. Gastroenterol. Hepatol.*, 1997, **9**(11), 1033–1044. P.W. Majerus, *Curr. Biol.*, 1998, **8**(3), R87–R89; (b) W. Gibb, *Ann. Med.*, 1998, **30**(3), 235–241.
64. D. Hwang, D. Scollard, J. Byrne, E.Levine, *J. Nat. Cancer Inst.*, 1998, **90**(6), 455–460; H. Vainio, *Br. Med. J.*, 1998, **317**(7162), 828–828.
65. T.H. Hida, J. Leyton, A.N. Makheja, P. Ben-Av, T. Hla, A. Martinez, J. Mulshine, S. Malkani, P. Chung, T.W. Moody, *Anticancer Res.*, 1998, **18**(2A), 775–782;T. Hida, Y. Yatabe, H. Achiwa, H. Muramatsu, K. Kozaki, S. Nakamura, M. Ogawa, T. Mitsudomi, T. Sugiura, T.Takahashi, *Cancer Res.*, 1998, **58**(17), 3761–3764.
66. J. Ghosh and C.E. Myers, *Nutrition*, 1998, **14**(1), 48–49.
67. (a) S.Y. Buckman, A. Gresham, P. Hale, G. Hruza, J. Anast, J. Masferrer, A.P. Pentland, *Carcinogenesis*, 1998, **19**(5), 723–729; (b) E. Kokoglu, Y. Tuter, K.S. Sandikci, Z. Yazici, E.Z. Ulakoglu, H. Sonmez, E. Ozyurt, *Cancer Lett.*, 1998, 132(1–2), 17–21.
68. S. Hashimoto, T. Kurokawa, T. Nonami, A. Harada, A. Nakao, H. Takagi, *J. Clin. Biochem. Nutr.*, 1998, **24**(1), 13–21.
69. E. Pennisi, *Science*, 1998, **280**(5367), 1191–1192. K.M. Prasad, *et al.*, *Proc. Soc. Exp. Biol. Med.*, 1998, **219**(2), 120–125.
70. A. Terao, H. Matsumura, M. Saito, *J. Neurosci.*, 1998, **18**(16), 6599–6607; K. Gerozissis, Z. deSaint Hilaire, A. Python, C. Rouch, M. Orosco, S. Nicolaidis, *Neuroreport*, 1998, **9**(7), 1327–1330.
71. R. Packer and B. Stanek, *Eur. Heart J.*, 1997, **18**(2), 318–329.
72. T.Z. Liu, A. Stern, J.D. Morrow, *J. Biomed. Sci.*, 1998, **5**(6), 415–420; M.J. Mueller, *Chem. Biol.*, 1998, **5**, R323–333.
73. W.E. Kaufmann, K.L. Andreasson, P.C. Isakson, P.F. Worley, *Prostaglandins*, 1997, **54**(3), 301–624.
74. D.S. DeWitt and D.S. Prough, *Crit. Care. Med.*, 1998, **26**(5), 819–821.
75. M.J. Norden, in *Beyond Prozac*. Regan Books (HarperCollins Publishers), New York, 1996.
76. T. Tsuri, *J. Med. Chem.*, 1997, **40**, 3504–3507.

9
The Importance of Mycobacterial Lipids

David E. Minnikin,[1] Michael R. Barer,[2] Angela M. Gernaey,[1,3] Natalie J. Garton,[1,2] James R. L. Colvine,[1] James D. Douglas[1] and Ali M. S. Ahmed[1]

[1] DEPARTMENT OF CHEMISTRY, UNIVERSITY OF NEWCASTLE, NEWCASTLE UPON TYNE NE1 7RU, UK

[2] SCHOOL OF MICROBIOLOGICAL, IMMUNOLOGICAL AND VIROLOGICAL SCIENCES, UNIVERSITY OF NEWCASTLE, NEWCASTLE UPON TYNE NE2 4HH, UK

[3] FOSSIL FUELS AND ENVIRONMENTAL GEOCHEMISTRY, UNIVERSITY OF NEWCASTLE, NEWCASTLE UPON TYNE NE1 7RU, UK

1 Introduction

Tuberculosis (TB) continues to kill some 3 million people a year, 95% of whom are in developing countries, and numbers are climbing.[1,2] Last year there were 8 million new cases and it is estimated that one third of the world's population—nearly 2 billion people—is infected with latent TB. The World Health Organisation (WHO) have also predicted that 30 million people will die from the disease over the next decade.[1,2]

The causative agent for TB is a member of the genus *Mycobacterium* known as *Mycobacterium tuberculosis*. Mycobacteria, in general, are considered to be amongst the oldest bacteria on earth and are ubiquitous within the environment.[3] It has been speculated that the source of human infection dates back to Neolithic times, perhaps when man first started to herd cattle. *Mycobacterium bovis* is another member of the 'tuberculosis complex,' infecting mainly cattle but having a broad host range including man. *M. tuberculosis*, however, is pathogenic only to man and it is possible that it might have originated from a mutation in *M. bovis*.

A major scientific break-through occurred in Berlin, in 1882, when Robert Koch (1843–1910) became the first person to culture the tubercle bacillus outside the body. After successfully inoculating the bacillus into an experimental animal, the practice of testing for the mycobacteria in the sputum of suspects became common. The generic name *Mycobacterium* was proposed by Lehmann and Neumann[4] in 1896 on the basis of the mould-like pellicles produced by mycobacteria when grown on a liquid medium. Within a couple of years, Theobold Smith identified microscopic, morphological and toxicological differences between various bacilli. Human, bovine, avian, murine and piscine varieties were all recognised, with the human bacillus accounting for 98% of pulmonary TB. The mycobacteria's most distinctive property is their unique staining. They are notoriously hard to stain and once stained resist decolorisation by acids and are thus known as acid-fast bacilli. Mycobacteria are non-motile, aerobic, slightly curved or straight rods and are 0.2–0.6 by 1.0–10 μm in size.

Members of the slow growers include *M. tuberculosis, Mycobacterium ulcerans* (responsible for the Buruli ulcer) and *Mycobacterium marinum* (a swimming pool granuloma). *Mycobacterium avium, Mycobacterium kansasii* and *Mycobacterium xenopi* are also slow growing mycobacteria that can lead to non-specific infections. These mycobacteria can either mimic pulmonary TB or can cause extrapulmonary infections such as abscesses. *Mycobacterium fortuitum*, whose most common clinical manifestation is an abscess, *Mycobacterium chelonae* and *Mycobacterium smegmatis* are all fast growing mycobacteria. Finally, *Mycobacterium leprae*, the causative agent for leprosy, is in a subclass of its own due to its inability to be cultured *in vitro*.

Initial infection with TB is usually the result of inhaling the bacilli contained in the sputum of someone with pulmonary TB. These bacilli, once in the lung, cause a small area of inflammation. This small area of inflammation provides a platform from which the bacilli can reach the lymph nodes at the base of the lung and enter the blood stream where they are free to reach many organs in the body. The infection is usually overcome by the body's immune system protecting the individual for life. However, if the victim's natural defences have been lowered through old age, chronic illness or alcoholism, for example, or they are immunocompromised then the primary infection may go on to cause progressive pulmonary TB. Infection with the Human Immunodeficiency Virus (HIV) is a prime cause of patients becoming immunocompromised and this undoubtedly plays a major role in the resurgence of TB. If HIV-negative persons become infected with TB, they face a 10% chance of developing the disease during their lifetime. By contrast, in any given year, an HIV-positive individual infected with TB faces an annual risk of developing progressive pulmonary TB as high as 8%. However, most cases of TB in patients with HIV infection are likely to result from reactivation of TB infection often acquired years before. There is also evidence to suggest that the mycobacteria increase HIV's rate of replication and accelerate the onset of full-blown AIDS.

2 The composition and Organisation of the Mycobacterial Cell Envelope

Mycobacteria have cell envelopes of unusually low permeability which contribute to their resistance to common antibiotics and chemotherapeutic agents.[5] Mycobacteria are also relatively resistant to drying, alkali and disinfectants, making it difficult to prevent the transmission of *M. tuberculosis*. These properties are thought to be related to the unique structure of the mycobacterial cell envelope. In order to investigate this prediction, a number of ultrastructural investigations have been carried out. Early work used electron microscopy and a combination of staining techniques. More recent investigations,[6] using freeze-substitution techniques, have produced similar conclusions. The cell wall is composed of an inner layer of moderate electron density, a wider electron-transparent layer and an outer electron-opaque layer of variable appearance and thickness. The inner layer's electron density pattern is consistent with a product containing carboxyl groups that can bind metal ions, the peptidoglycan layer. The electron-transparent layer appears to be the hydrophobic barrier dominated by lipids covalently bound to the arabinogalactan. The nature of the outer most electron-opaque layer is less well understood due to variations in thickness, electron density and appearance. However, ruthenium red dye[7] has allowed consistent visualisation, suggesting a structure containing negatively charged head groups of lipids. Additional carbohydrate material and proteins have also been identified.

The mycobacterial plasma membrane is homologous to the plasma membranes of other bacteria.[8,9] Polar lipids, mainly phospholipids, assemble themselves in association with proteins into a lipid bi-layer. It is also assumed that most mycobacterial plasma membranes are similar in terms of lipid composition.

Mycobacterial peptidogycan consists of alternating units of *N*-acetyl glucosamine linked $\beta 1 \rightarrow 4$ to muramic acid (see Figure 1). Within the cell wall there

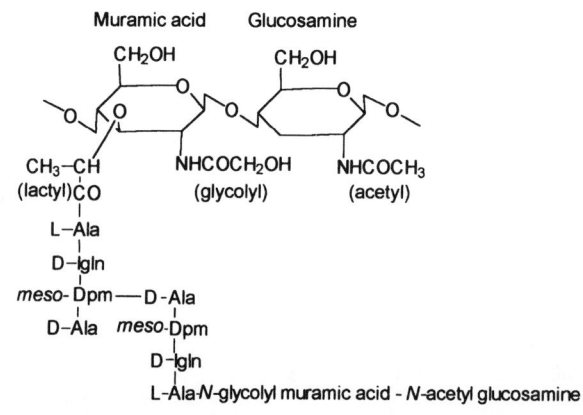

Figure 1 *Essential structure of mycobacterial peptidoglycan. Abbreviations: Ala, alanine; Igln, isoglutamine; Dpm, diaminopimelic acid; Gly, glycine*

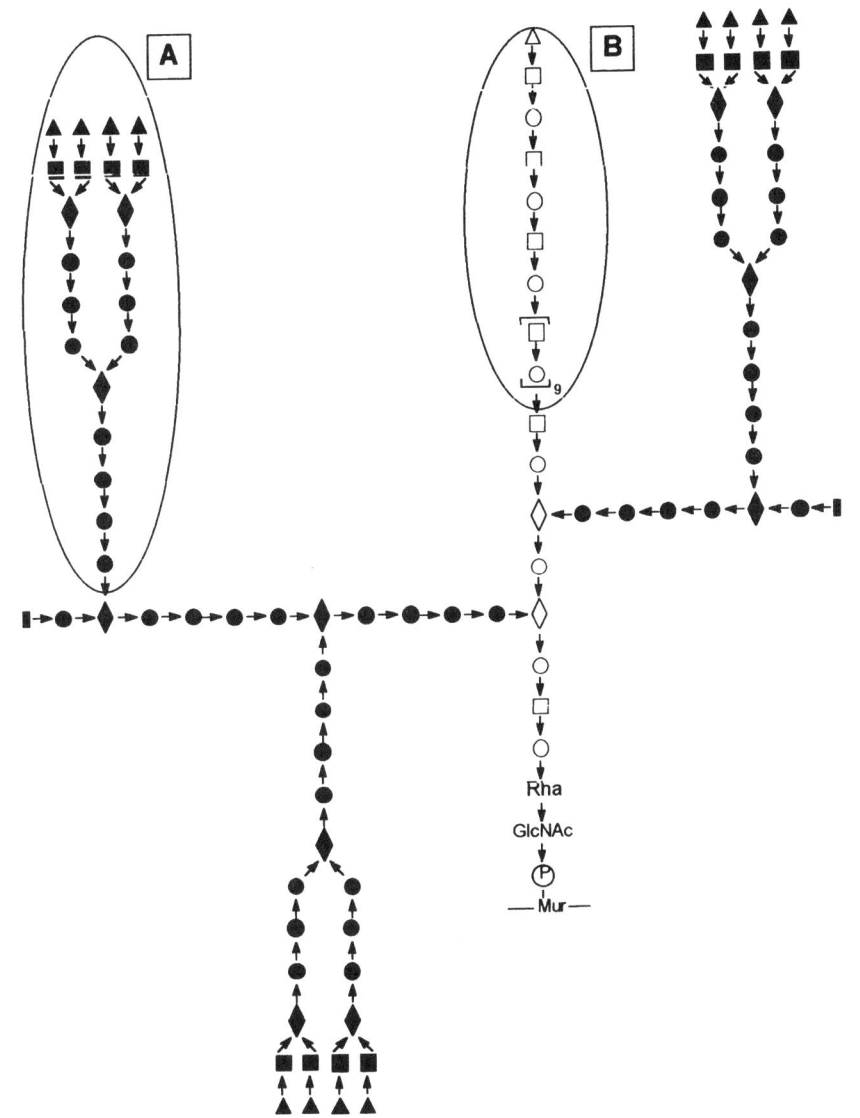

●, 5-α-D-Ara$f$; ◆, 3,5-α-D-Ara$f$; I, t-α-D-Ara$f$; ■, 2-α-D-Ara$f$ or 2-α-D-Ara$f$ mycolyl substituted;

▲, t-β-D-Ara$f$ mycolyl substituted; △, t-β-D-Gal$f$; □, 6-β-D-Gal$f$; ○, 5-β-D-Gal$f$; ◇, 5,6-β-D-Gal$f$.

Figure 2 *Proposed structure of the AG-linkage unit-peptidoglycan of the mycobacterial cell wall. A and B represent the extended non-reducing end of the arabinan and a large proportion of the galactan chain, respectively*

is also a unique arabinogalactan polysaccharide (Figure 2).[5,10,11] Arabinogalactan is composed of three arabinan chains (composed of about 27 D-arabinofuranosyl units each) attached to the homogalactan core (consisting of 32 D-galactofuranosyl units). The galactan is terminated by a linker oligosaccharide attached to the peptidoglycan through a phosphodiester bond.

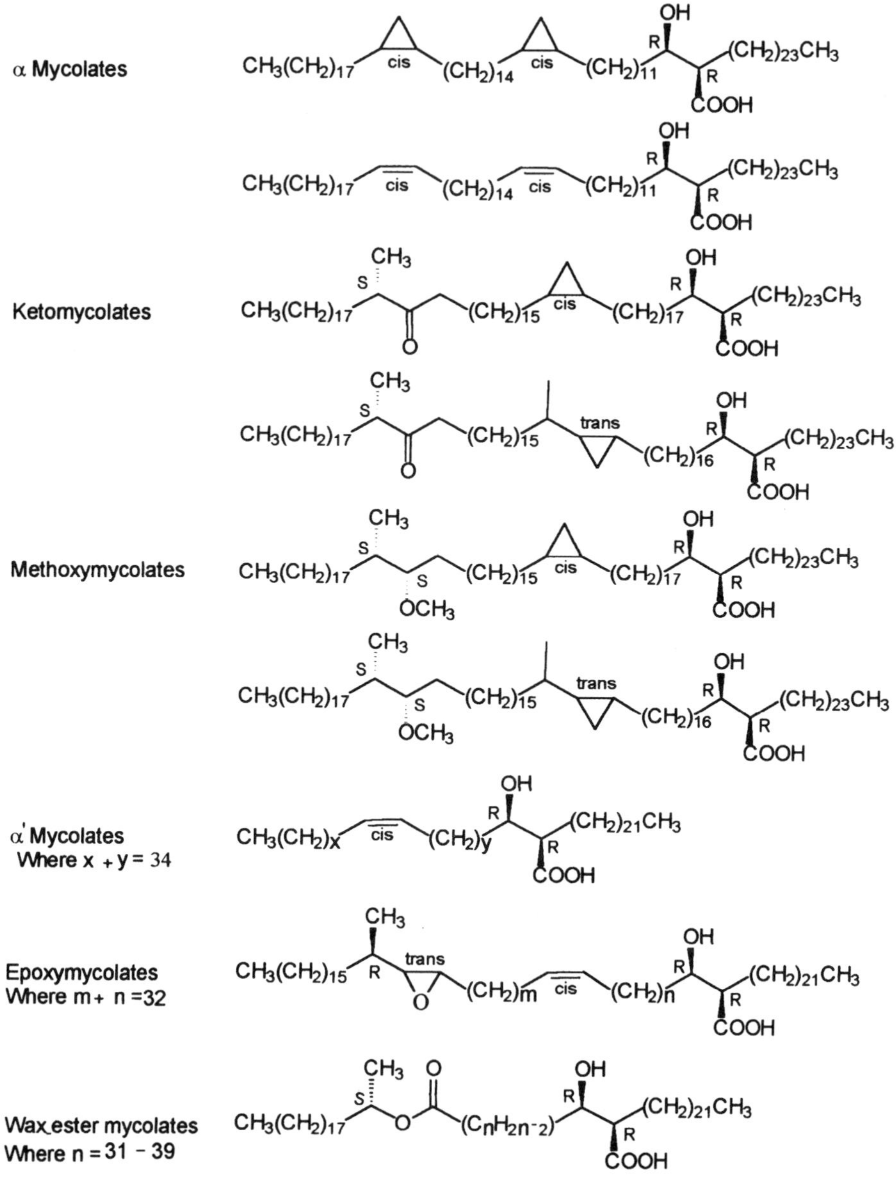

Figure 3 *Structures of representative mycobacterial mycolic acids*

Roughly two thirds of the terminal pentaarabinosyl units of the polysaccharide are esterified by long-chain mycolic acids, whose structures will be introduced later (Figure 3). All the galactosyl residues within the arabinogalactan are in the furanose form. The galactan, as mentioned previously, exists as a homopolysaccharide consisting of alternating 5- and 6-linked β-D-Gal*f* residues. Approximately 12 units of this repeat disaccharide comprise the entire homogalactan segment. There is also evidence to suggest that two arabinan chains are attached to the galactan back-bone, each at the C-5 position of 6-linked Gal*f* residues. Linear 5-linked α-D-Ara*f* residues with branching introduced *via* 3,5-linked α-D-Ara*f* units substituted at both branch positions with 5-linked α-D-Ara*f* residues comprise the majority of the homoarabinan chains. Finally, a non-reducing arabinan terminal pentasaccharide, or a pentaarabinosyl, has been encountered.

Mycolic acids were first identified in an unsaponifiable lipid extract isolated from *M. tuberculosis* by Stodola *et al.* (1938).[12] They are also found in the genera *Corynebacterium, Gordona, Mycobacterium, Nocardia, Rhodococcus* and *Tsukamurella*, comprising the so-called 'mycolata' branch of the actinomycete line. Mycobacterial mycolic acids differ from those found in 'mycolata' genera by the following features:

a) They are the largest mycolic acids (C_{70} to C_{90}),
b) They have the largest side chain (C_{20} to C_{24}),
c) They contain one or two unsaturations (in the form of double-bonds or cyclopropane rings),
d) They contain additional oxygen functionalities to the β-hydroxy group,
e) They have methyl branches in the main carbon backbone.

Key structural features, such as methoxy, keto and methyl branches, double-bonds and cyclopropane rings, are consistent with the main-chain portion; conservation of the α-branch has also been established with the mycolic acid family. However, various categories of mycolic acids have been well defined. The α-mycolates are those mycolic acids that lack any extra oxygen functionality in addition to the β-hydroxy acid group. The α-mycolates isolated from virulent *M. tuberculosis* have been shown to contain one major series containing two *cis*-cyclopropane rings. In addition keto- and methoxy-mycolates have been characterised. The majority of ketomycolates are composed of a series containing *trans*-cyclopropane rings compared to *cis*-cyclopropane rings found in the majority of methoxymycolates.

The nature of the free lipids encountered within the mycobacterial cell envelope varies greatly, particularly the differing polarities. The least polar lipids are the waxes, such as the dimycocerosates of the phthiocerol family, which consist of multimethyl branched mycocerosic acids esterified to long-chain diols (Table 1).[8,13] Mycocerosic acids are composed of a complex mixture of multimethyl branched fatty acids with a major C_{32} component, 2,4,6,8-tetramethyloctacosanoate (Table 1).

Phenolphthiocerols are also classed within the phthiocerol family and they form the basic lipid core of phenolic glycolipids. Phenolic glycolipids are most commonly encountered in a limited number of slow-growing mycobacteria[5,8,10,13] and consist of a conserved lipid core with a carbohydrate moiety, linked to the phenol unit. The lipid core is composed of long-chain diols esterified by multimethyl branched fatty acids. The carbohydrate part consists of *O*-methylated deoxysugars and small oligosaccharides.

A unique family of glycoconjugates, the glycopeptidolipids, have also been

Table 1 *Dimycocerosates of the phthiocerol and phenolphthiocerol families*

Phthiocerol A	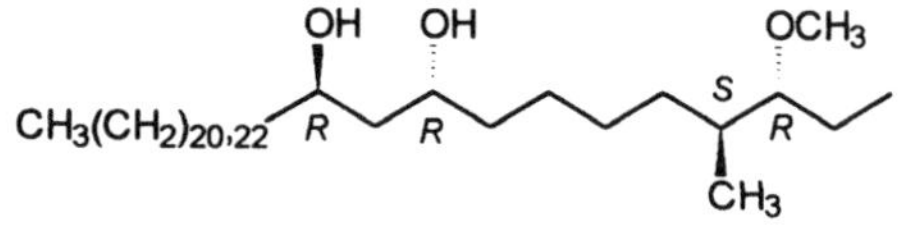 CH₃(CH₂)₂₀,₂₂–R–CH(OH)–R–CH(OH)–R–(chain)–R–CH(OCH₃)–CH(CH₃)–R
Phthiocerol B	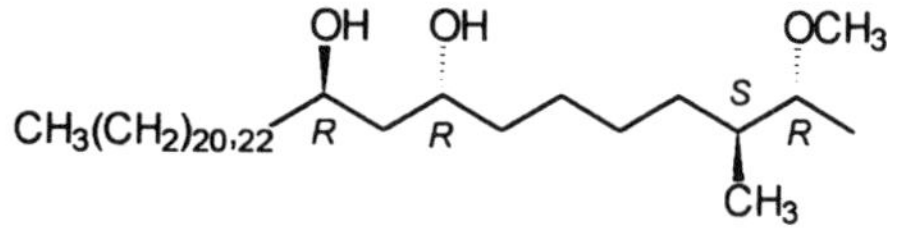 CH₃(CH₂)₂₀,₂₂–R–CH(OH)–R–CH(OH)–R–(chain)–R–CH(OCH₃)–CH(CH₃)–R
Phthiodiolone A	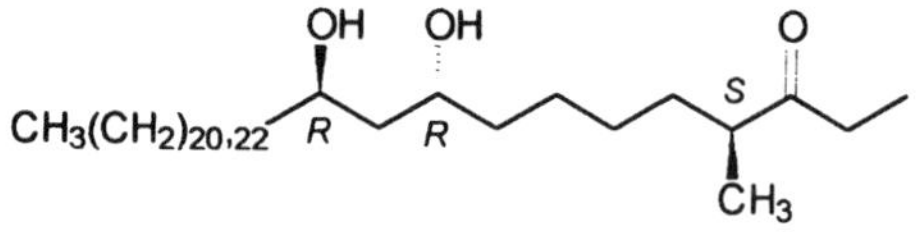 CH₃(CH₂)₂₀,₂₂–R–CH(OH)–R–CH(OH)–R–(chain)–R–C(=O)–CH(CH₃)–R
Phthiotriol A	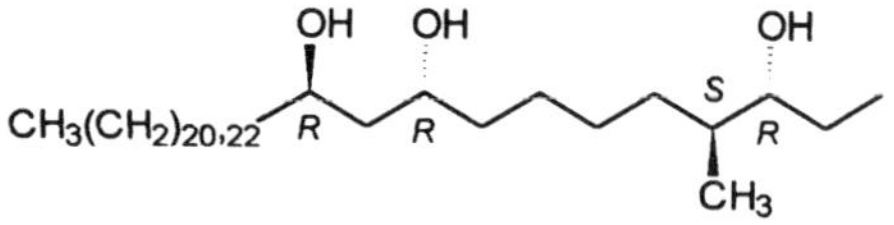 CH₃(CH₂)₂₀,₂₂–R–CH(OH)–R–CH(OH)–R–(chain)–R–CH(OH)–CH(CH₃)–R
Phenolphthiocerol A	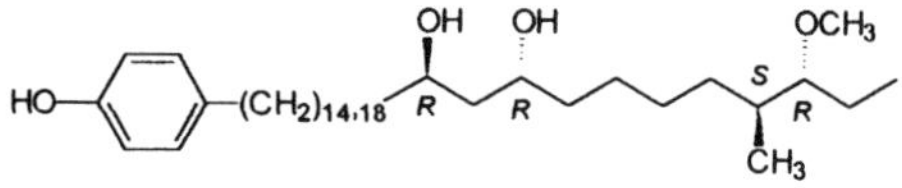 HO–C₆H₄–(CH₂)₁₄,₁₈–R–CH(OH)–R–CH(OH)–R–(chain)–R–CH(OCH₃)–CH(CH₃)–R
Mycocerosic acid	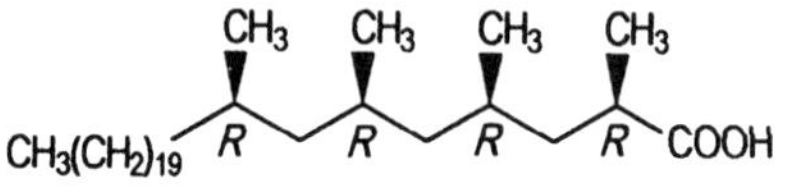 CH₃(CH₂)₁₉–R–CH(CH₃)–R–CH(CH₃)–R–CH(CH₃)–R–CH(CH₃)–COOH

encountered in isolates from both slow- and fast-growing mycobacteria[5,10] (see Figure 4). The lipid component of these glycosylated lipopeptides is a mixture of 3-hydroxy and 3-methoxy long-chain fatty acids amidated by a tripeptide terminated by an amino alcohol (L-alaninol). The alcohol is glycosylated by an O-methylated mono- or di-rhamnosyl residue. Additional sugar units, increasing the polarity of the lipids, have also been encountered.

Several families of lipids are based on trehalose such as acylated trehaloses known as 'cord factor' (Figure 5) and the sulfatides (Figure 6). Cord factor was first isolated from *M. tuberculosis* by Bloch in the 1950s and was

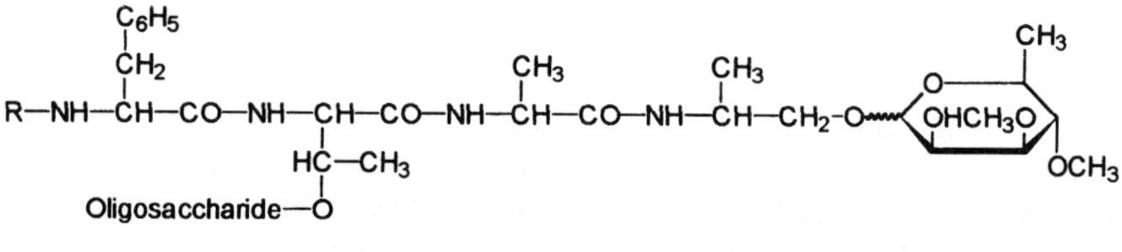

Figure 4 *Example of a glycopeptidolipid where R = long chain fatty acyl*

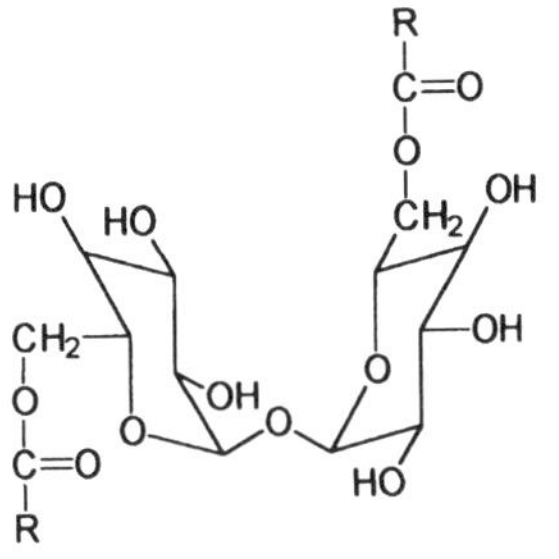

Figure 5 Trehalose 6,6'-dimycolate

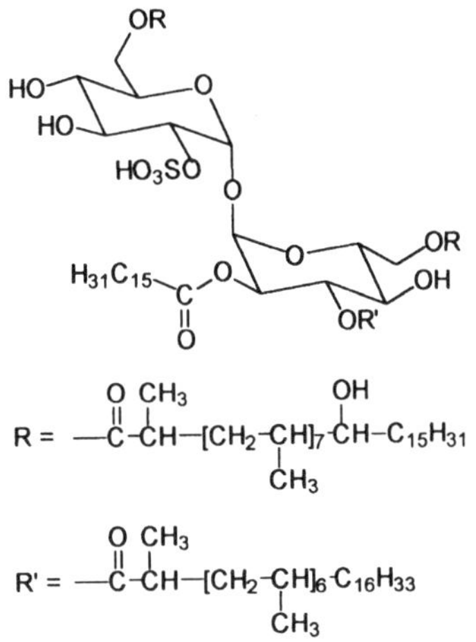

Figure 6 2,3,6,6'-Tetra-O-acyltrehalose 2'-sulfate

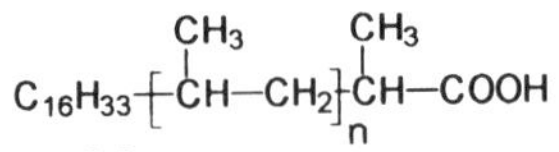

Figure 7 Phthioceranate

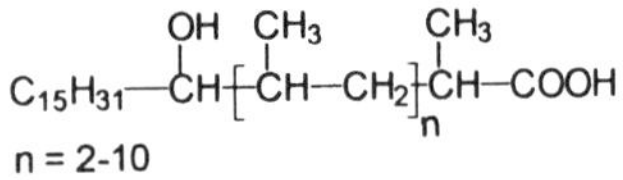

Figure 8 Hydroxyphthioceranate

subsequently identified as trehalose 6,6'-dimycolate.[10] The sulfatides are based on trehalose 2'-sulfate acylated in various positions with phthioceranate (Figure 7), hydroxyphthioceranate (Figure 8) and saturated straight-chain fatty acids. Despite a lack of evidence to support either the cord factor's or the sulfatide's role in cord formation, they are still implicated in a number of key disease-related events.[10]

A model for the possible interaction of covalently bound mycolic acids and complex free lipids was first elaborated by Minnikin[8] in 1982. An updated version of this model (Figure 9) shows how the typically flexible plasma membrane relies on the peptidoglycan–arabinogalactan complex to provide the cell with shape and integrity. The mycolic acids are covalently bound to the arabinogalactan, possibly in a parallel arrangement. This structural feature allows the mycolic acids to pack together forming a lipid monolayer that forms a hydrophobic barrier. Complex free lipids, such as phenolic glycolipids and phthiocerol dimycocerosates, can then plug any gaps that may exist if the mycolate arrangement is not totally uniform.

Mycolic acids are mainly found in clusters of four as covalently bound esters of arabinogalactan with the remainder present in lipids that can be extracted with organic solvents (Figure 10).[5,10,11]

Compelling evidence has been obtained recently[14] in support of the proposal that two lipid domains are present in the mycobacterial cell envelope. Treatment of cells with the fluorescent probes, dodecanoylaminofluorescein and hexadecanoylaminofluorescein, showed that the former agent was distrib-

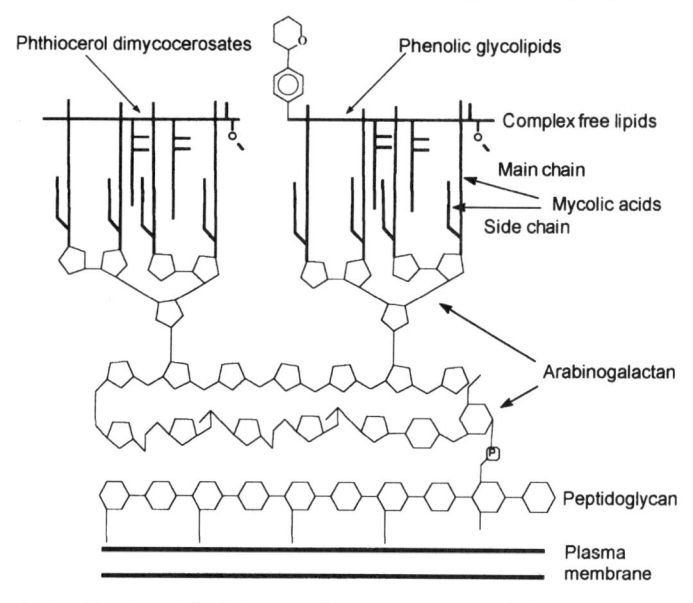

Figure 9 *A simplified model of the possible arrangement of chemical components in the mycobacterial cell envelope*

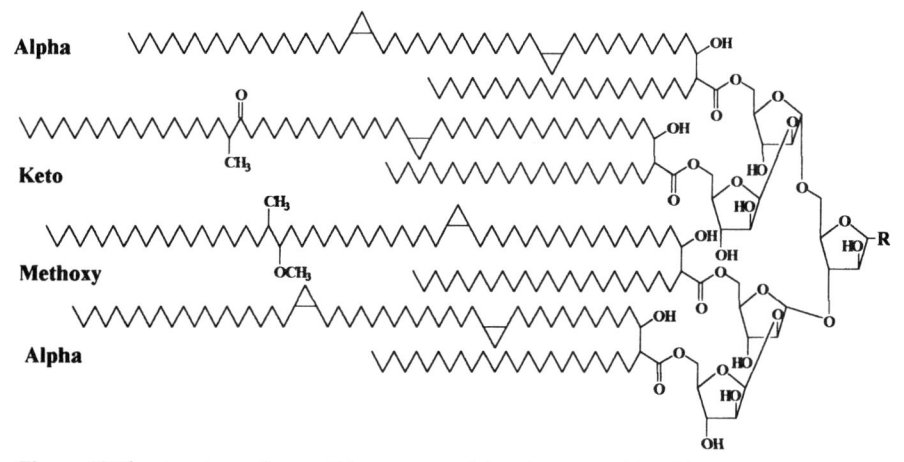

Figure 10 *The structure of a possible tetramycolylated pentaarabinoside.*

uted throughout the envelope but the more lipophilic latter probe was restricted in location to regions external to the plasma membrane.[14] This clearly suggests that a mycolic acid monolayer (Figure 10) interacts with other lipids to produce an outer lipid domain as depicted in Figure 9.

3 Current Anti-tuberculosis Agents

Active TB is currently treated with a combination of drugs: isoniazid (**1**) (INH), rifampicin (**2**), pyranizamide (**3**) and ethambutol (**4**) (see Figure 11).[15]

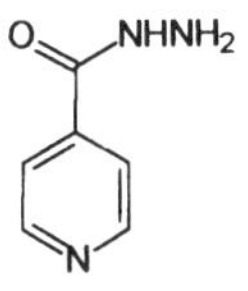

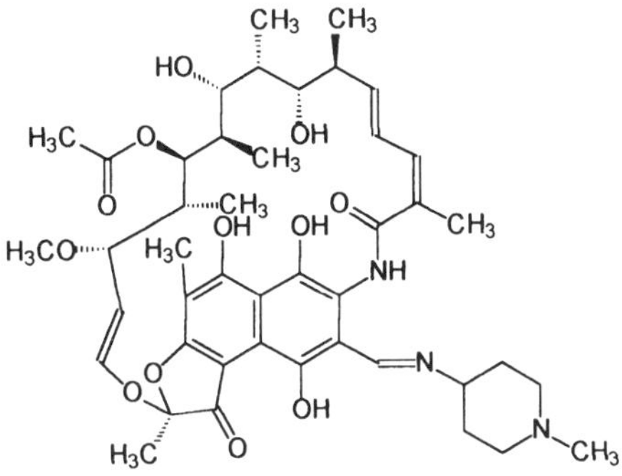

Isoniazid (**1**)

Rifampicin (**2**)

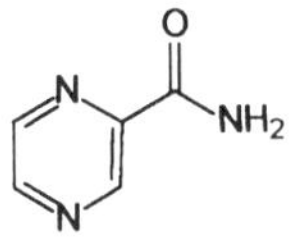

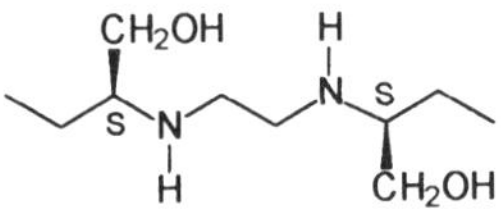

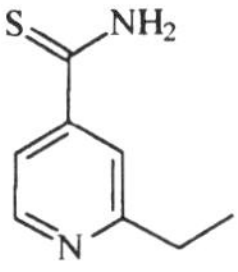

Pyrazinamide (**3**)

Ethambutol (**4**)

Ethionamide (**5**)

Figure 11 *Structures of some antimycobacterial compounds*

The patient is initially treated with INH, rifampicin and pyrazinamide for 2 months, with a continuation of INH and rifampicin for 6 months. This treatment is then supplemented with ethambutol, providing an effective regimen against susceptible organisms in pulmonary infections. However, a combination of patients failing to complete the therapy and inappropriate use of the drugs individually has led to the emergence of Multi-Drug Resistant (MDR) strains of TB. In order to progress with the research into the development of superior drugs, the current treatments need to be closely scrutinised.

The preferred solution to dealing with an infectious disease is the use of preventative medicine, *i.e.* vaccination. An estimated 100 million people are vaccinated against TB every year.[1] Patients are vaccinated with a live attenuated strain of *M. bovis* referred to as the bacille Calmette-Géurin (BCG). Although effective, BCG has aroused controversy since its introduction in the 1950s. Clinical trials have highlighted inconsistencies in the efficacy of the vaccination. In the United Kingdom, for example, BCG offered greater than 80% protection compared to 0% in parts of southern India[1]. A commonly accepted explanation for this observation is that non-pathogenic mycobacteria, ubiquitous within the environment, mask the effect of the vaccination. However, BCG remains of value due to the protection it provides against childhood TB, meningitis and primary disseminated TB.

INH (isonicotinic acid hydrazide) (**1**) was first reported to be effective in the treatment of *M. tuberculosis* in 1952.[16] Both *M. tuberculosis* and *M. bovis* BCG are extremely susceptible to the drug in the range of 0.0–0.2 μg/mL. INH is bactericidal and is the most commonly prescribed drug for treatment and prophylaxis. INH shows poor antibacterial activity against other bacteria and was initially thought to harbour a low risk of harmful side effects. However, 1 in 7000 people on a 6 month course of INH die of liver damage. which has encouraged European doctors to limit the use of INH to high-risk patients. It was demonstrated[17] that INH inhibits the synthesis of mycolic acids in *M. tuberculosis.* A specific inhibitory effect was observed on the synthesis of saturated fatty acids greater than 26 carbons[18] and the synthesis of mono-unsaturated fatty acids *in vivo*.[19] Further research indicated that INH partially inhibited a very long-chain fatty acid elongation system in *M. avium*.[14] The implications were that an enzymatic step, or steps, in the elongation of fatty acids was the site of action of INH. It has since been demonstrated that INH is actively taken up by *M. tuberculosis,* and is oxidised by the mycobacterial catalase-peroxidase[20,21] to an electrophilic species. This activated form of INH binds to, and inhibits the activity of, the *M. tuberculosis* enoyl–ACP reductase–NADH complex essential for fatty acid elongation.[22,23] In addition, INH is a potent inhibitor of the peroxidatic reaction catalysed by the *M. smegmatis* catalase-peroxidase.[24] The related drug, ethionamide (5) (see Figure 11) is also metabolised to an electrophilic intermediate but by a different mycobacterial enzyme from INH.[23] Once activated, both these drugs irreversibly inactivate the enoyl–ACP reductase–NADH complex required for fatty acid elongation.

The primary use of rifampicin (2) is to reduce the duration of therapy from 18 months to 9 months. It works by inhibiting the β-subunit of the DNA-dependent RNA polymerase encoded by the *rpo*B gene. MDR strains of *M. tuberculosis* and *M. smegmatis* and *M. leprae* mutants, resistant to rifampicin, are mutated in this gene.[25–28]

The mode of action of pyranizamide (3) is analogous to that of INH in that it is thought to be a pro-drug that is deamidated by pyrazinamidase to form the free acid. A noticeable difference, however, is the inactivity of pyrazinamide at alkaline pH. Confirmation of mode of action was achieved by cloning the gene from MDR TB which encodes this enzyme (*pnc*A). Mutated *pnc*A gene was isolated from clinical isolates resistant to pyrazinamide correlating with a loss of pyranizamidase activity. Replacing mutant *pnc*A with wild-type *pnc*A restored activity.[29]

Takayama and Kilburn[30] observed that arabinose was incorporated into the cell wall arabinogalactan. Further studies showed that arabinose incorporation into the cell wall was specifically inhibited by ethambutol (4), while incorporation in lipoarabinomannan was not as severely affected. Mutants of *M. smegmatis* resistant to ethambutol showed that the effects did extend to lipoarabinomannan although not to such an extent as that seen in arabinogalactan. The effect of ethambutol on arabinose transfer was clearly demonstrated by using a chemically synthesised polyprenol phosphate donor in cell-free assays.[31]

4 Mycolic Acid Biosynthesis as a Drug Target

The mycolic acids appear to be an integral part of the mycobacterial cell envelope and their unique structures suggest that their biosynthesis may be a valuable target for drugs. Early investigations into the biosynthesis of mycolic acids indicated that C_{32} mycolic acids are formed by a Claisen-type condensation and reduction of C_{16} fatty acid in *Corynebacterium diphtheriae*. Similar biosynthetic pathways were established in *Nocardia asteroides* and *M. smegmatis*. The proposed pathway[32] (Figure 12) involves four distinct stages;

1. Synthesis of C_{24}–C_{26} straight-chain saturated fatty acids providing C-1 and C-2 atoms and the α-alkyl chain
2. Synthesis of C_{40}–C_{60} acids (meromycolic acids) providing the main carbon backbone, followed by elongation
3. Functional group introduction *e.g.* Δ^5-desaturase enzyme acting on a C_{24} acid
4. Condensation reaction affording the mycolic acid

The syntheses of C_{24} and C_{26} fatty acid primers are catalysed by fatty acid synthases (FASs) and elongases, specifically FAS-I and FAS-II complexes. FAS-I is a single polypeptide with multiple catalytic activities that generate several shorter CoA esters from the acetyl CoA primer.[33–36] The primary products of the *de novo* FAS-I system are C_{16} to C_{18} and C_{24} to C_{26} fatty acyl-CoA derivatives creating precursors for further elongation.[32,37] FAS-II is incapable of *de novo* fatty acid synthesis but instead elongates a C_{16} fatty acid primer (palmitoyl-ACP) to fatty acids ranging from C_{24} to C_{56} carbons in length.[38] In *M. tuberculosis* these functions are linked in a multifunctional enzyme complex that appears to be a *de novo* synthase joined to an elongase that lengthens C_{24} or C_{26}.[39] However, it is still unclear if FAS-II or an elongase enzyme is responsible for the synthesis of meromycolates from C_{24} fatty acids. A key step in the biosynthesis appears to be the Δ^5-desaturation of tetracosanoate (24:0) to 24:1 *cis*-5.[40] Elongation of 24:1 *cis*-5 is possibly followed by the action of a Δ^3-desaturase, providing a second point of unsaturation. Further elongation and cyclopropanation can provide the so-called meromycolate which can be condensed with tetracosanoate to give mycolic acids. Details of the biosynthesis of oxygenated mycolic acids are less well developed.

(Z)-Tetracos-5-enoic acid (**6**) is a key intermediate in the biosynthesis of mycobacterial mycolic acids[41] and a structural analogue, 1-(3'-methoxycarbonylpropyl)-2-(octadecyl)cyclopropene (**7**), has been shown to inhibit mycolate synthesising activity.[42] This result was the first of an entirely novel approach to developing antimycobacterial chemotherapeutic agents whose

$CH_3(CH_2)_{17}$ —— $(CH_2)_3CO_2H$ $CH_3(CH_2)_{17}$ △ $(CH_2)_3CO_2CH_3$

(**6**)   (**7**)

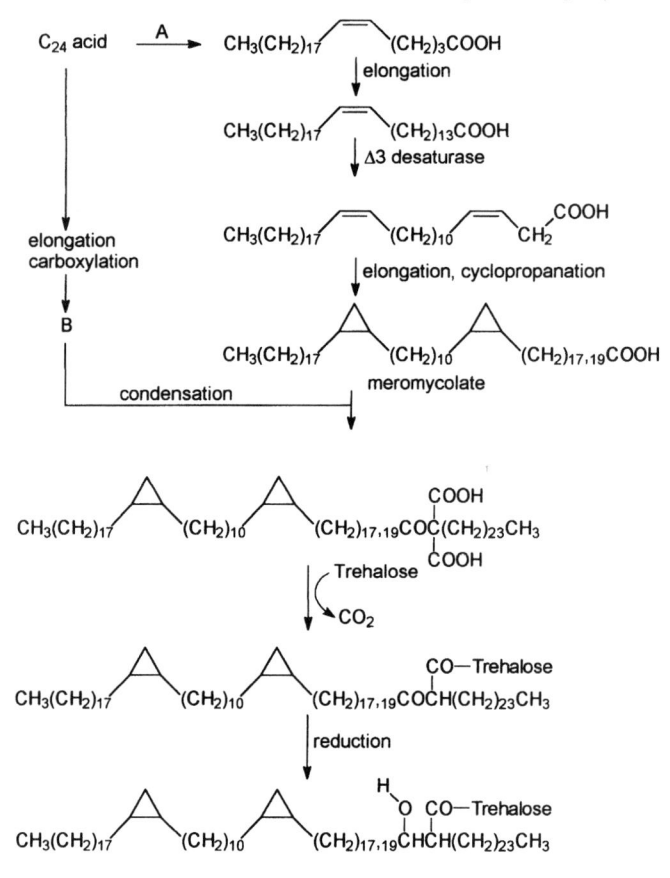

Figure 12 *Biosynthesis of α-mycolic acids in* M. *tuberculosis H37Ra. A, Δ^5-desaturase; B, tetracosanyl malonate, $CH_3(CH_2)_{23}CH(COOH)_2$*

specific target is cell wall biosynthesis. 1-(3′-Methoxycarbonylpropyl)-2-(octadecyl)cyclopropene was active at 1–2 mg/mL and was one of the first compounds to be tested as a methyl ester. This was made possible by the use of a now well-established hexane–water mycolate synthesising system which includes cell wall material.[41] It is presumed that the preparation contains lipases to liberate the free acids, allowing the use of methyl esters.

In 1982, the structure and antibiotic properties of thiolactomycin (**8**) (TLM) were first reported by Sasaki *et al.*[43] TLM, (4R)-(5E)-2,4,6-Trimethyl-3-hydroxy-2,5,7-octatriene-4-thiolide, was isolated from a soil sample containing an organism of the genus *Nocardia* and was the first example of a naturally occurring thiolactone to exhibit antibiotic activity. The compound has mod-

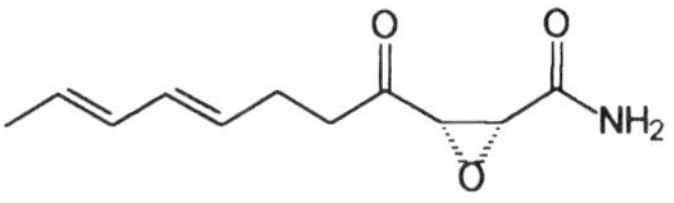

Cerulenin (**9**)

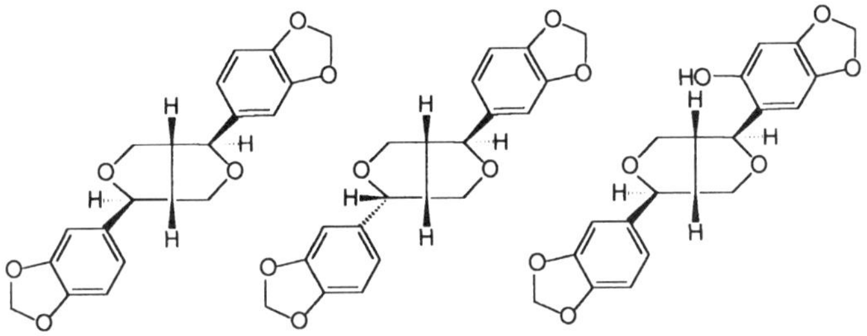

Sesamin (**10**) Episesamin Sesaminol

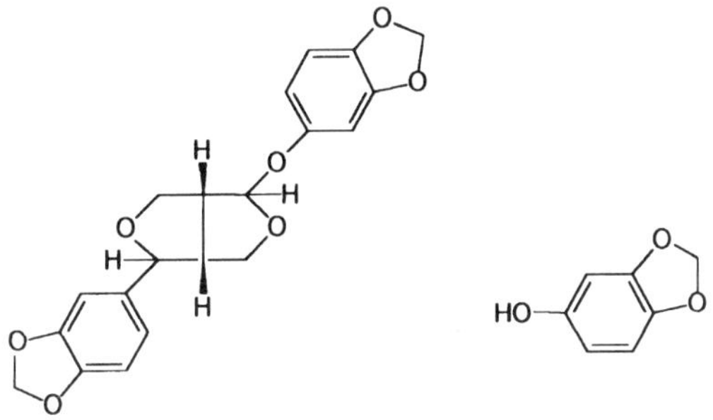

Sesamolin Sesamol

Figure 13 *Cerulenin, sesamin and sesamin related lignan compounds*

erate *in vitro* activity against a broad spectrum of pathogens including Gram-positive cocci, enteric bacteria, acid-fast bacteria and anaerobic bacteria. TLM has also shown encouraging anti-malarial activity[44] *via* the inhibition of the type-II fatty acid biosynthetic pathway in apicoplasts. Apicoplasts, or apicomplexan plastids, are responsible for a number of diseases including malaria and toxoplasmosis.

Cerulenin (**9**),[45] sesamin (**10**) and sesamin related lignan compounds (see Figure 13)[46] are further examples of natural products that inhibit fatty acid biosynthesis. Cerulenin, (2*R*3*S*)-2,3-epoxy-4-oxo-7,9-dodecadienolylamide, is a fungal product that is extremely effective at blocking the growth of a broad spectrum of bacteria and blocks β-ketoacyl-ACP synthase activity by irreversible, covalent modification of the synthase active site.[45] (+)-Sesamin, by contrast, is a non-competitive inhibitor of Δ^5-desaturase.[46] However, neither cerulenin or (+)-sesamin are useful antibiotics owing to lack of selectivity. Cerulenin inhibits the condensing enzyme reaction catalysed by the multifunctional mammalian (type I) fatty acid synthase and (+)-sesamin inhibits Δ^5-desaturase in polyunsaturated fatty acid biosynthesis in both microorganisms and animals. TLM, however, is potentially of chemotherapeutic value as it is non-toxic to mice and affords significant protection against urinary tract and intraperitoneal bacterial infections.

Early studies on the mode of action of TLM were executed on fatty acid synthesis in peas.[47] The use of [1-^{14}C]acetate labelling in pea chloroplasts revealed that TLM is an inhibitor of type II dissociable fatty acid synthases. The three β-ketoacyl acyl carrier protein (ACP) synthases were all inhibited by TLM to varying degrees. Condensing enzyme II, KAS-II, (which catalyses the elongation of palmitoyl-ACP to stearoyl-ACP), is the most sensitive, with the short-chain condensing enzyme, KAS-III (which catalyses the condensation of acetyl-CoA with malonyl-CoA), being the least sensitive. Further biological evaluation of TLM indicated *in vivo* antimycobacterial activity against saprophytic *M. smegmatis* and the virulent strain *M. tuberculosis* Erdman.[48] Through the use of *in vivo* [1,2-^{14}C]acetate labelling of *M. smegmatis*, TLM was shown to inhibit the biosynthesis of both fatty acids and mycolic acids. Synthesis of the shorter chain α'-mycolates of *M. smegmatis* was not affected but TLM did inhibit the synthesis of the longer chain α-mycolates and epoxymycolates (Figure 14).

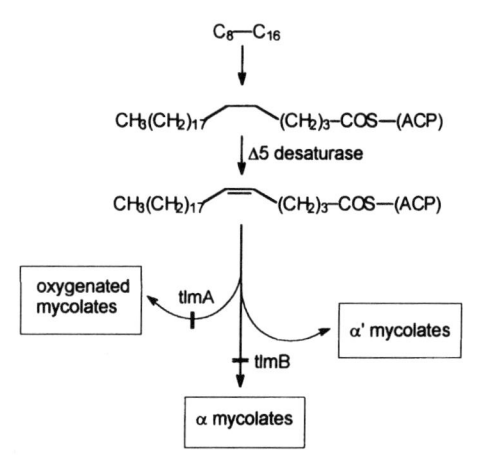

Figure 14 *TLM and mycolic acid biosynthesis. The sites of action of TLM are represented by tlmA and tlmB. The oxygenated mycolates include epoxy, keto, methoxy and wax ester mycolates typical of* M. smegmatis, M. tuberculosis *and* M. avium. *The α'-mycolates are representative of the monounsaturated mycolates of* M. smegmatis

The use of *M. smegmatis* cell extracts demonstrated that TLM specifically inhibited the mycobacterial acyl carrier protein-dependent type II fatty acid synthase (FAS-II) but not the multifunctional type I fatty acid synthase (FAS-I). Analysis of the *in vivo* and *in vitro* data has suggested two separate sites of action for TLM: the β-ketoacyl-acyl carrier protein synthase in FAS-II and the elongation step involved in the synthesis of α- and oxygenated mycolates. The possible characterisation of fatty acid and mycolic acid biosynthetic genes should reveal a novel range of chemotherapeutic agents directed against *M. tuberculosis*. Figure 15 shows both TLM's and INH's proposed sites of action.

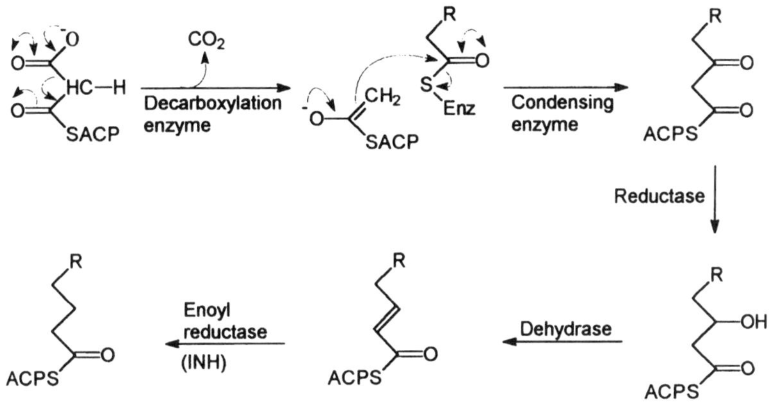

Figure 15 *Targets in mycobacterial fatty acid elongation*

Serendipity has always had its place in science and it played a major role in the discovery of isoxyl (ISO) (**11**) [1,3-bis-(*p*-3-methylbutoxyphenyl)thiourea] and related arylthiourea compounds. During the screening of a range of compounds, 1,3-bis-(*p*-ethoxyphenyl)thiourea (**12**) (see Figure 16) was shown to have high anti-tuberculosis activity in mice infected with the H37Ra strain.[49] Further investigations by Buu-Hoï and co-workers revealed that a range of 1,3-bis-(*p*-alkoxyphenyl)thioureas were active against human leprosy and also that thiourea (**13**) (see Figure 16) itself had moderate *in vitro* activity.[50–52] Appropriate substitution enhances activity further with allylthiourea (**14**) and *p*-aminobenzenesulfonylthiourea (**15**) (see Figure 16) showing superior activities. Despite very little being known about their mode of action, these compounds were used in a chemotherapeutic capacity during the 1950s.

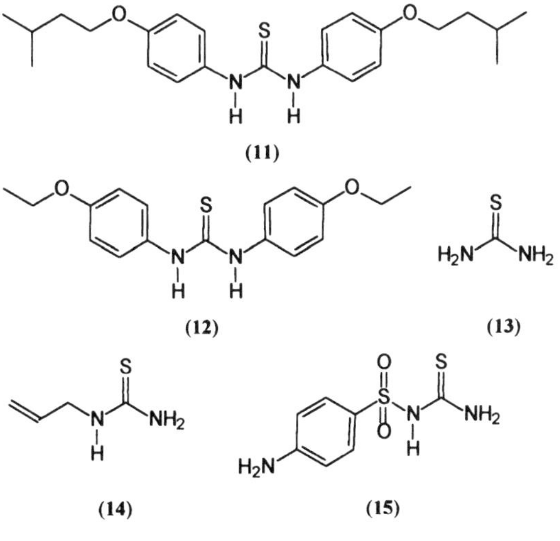

Figure 16 *Agents based on thiourea, with antimycobacterial activity*

Initial experiments showed that the tuberculostatic activity of ISO (**11**) is comparable with that of ethionamide (**5**) and no toxic side-effects were demonstrable.[53] In a recent reinvestigation of the activity of ISO,[54] it was found that mycolic acid synthesis is inhibited by this drug and a range of analogues.

5 Mycolic Acids as Biomarkers for the Detection of Ancient Tuberculosis

Human tuberculosis (TB) is currently on the increase in western populations. This has been globally linked to an increasing drug resistance and a malevolent synergy with HIV infection.[55] The disease is on the increase even in individuals with apparently competent immune systems.[56] If cyclical changes in the epidemiology of tuberculosis underlie these developments,[57] then the study of the disease in past populations may be of value in understanding this rise. TB is believed to be of great antiquity[58] as many ancient civilisations have described and depicted the disease.[3]

Studies of TB in past populations will be of value if reliable diagnosis can be made from archaeological human remains. Currently, diagnosis on the basis of osteological change is the accepted method, since TB induces bony changes in 1%[59] to 5%[60] of modern untreated adult cases, with no underlying acquired immunosuppressive disease. It is uncertain, however, whether any extrapolations may be made concerning disease frequency from the prevalence of these changes in ancient populations to the ancient population as a whole. When such extrapolations are made, the disease frequencies are always much less than expected from historical sources.[61]

The use of chemical biomarkers for diagnosis is potentially much more reliable than macroscopic examination of skeletal remains. Use of the polymerase chain reaction (PCR) offers the possibility of amplifying specific DNA targets, such as the insertion element *IS6110*, carried by most modern *M. tuberculosis* (MTB) isolates.[62] There have been recent reports of the PCR amplification of MTB-specific insertion element DNA sequences from ancient human bones[63] and mummified soft tissues[64] and gene sequencing of amplified products from soft tissue.[65]

Long-chain compounds, such as the mycolic acids (Figure 3) and phthiocerol dimycocerosates (Table 1) have the potential for long-term survival in infected material. Chemical and chromatographic methods are available for the sensitive detection of such materials,[66] possibly providing a more robust alternative to PCR amplification.

In an initial study (A.M. Gernaey, D.E. Minnikin, J.C. Middleton, R.A. Dixon and C.A. Roberts, unpublished) the presence of both MTB mycolates and DNA fragments was confirmed in a rib from a 1000 year old individual with Pott's disease. The rib had no osteological changes consistent with TB.[67] In another study, evidence was obtained for the presence of mycolic acids in calcified pleural material from remains 1400 years old.[68] These very promising initial results were substantiated by the analysis of a coherent set of archaeological individuals from a site with records detailing the incidence of tuberculosis.[69] The construction of the "International Centre for Life" Genetics Institute in Newcastle upon Tyne, timed to celebrate the Millennium, required the excavation of the first Newcastle Infirmary (AD 1753–1906) and its associated burial ground (AD 1753–1845). From November 1996 until March 1997, skeletal remains from 210 articulated individuals and a minimum of 407 dis-articulated individuals were recovered from the burial ground (J. Nolan, unpublished). The Infirmary Charter precluded the admission of those suffering from tuberculosis and children less than fifteen years old, but the skeletal remains of some children were recovered. The surviving Newcastle Infirmary records, for 1803 to 1845, showed that 27.1% of patients died with tuberculosis described by a range of pseudonyms (J. Nolan, unpublished). The hospital functioned at a time when tuberculosis was seen as a major cause of death in Britain,[70] but osteological examination has shown only 2/210 (0.95%) individuals with skeletal changes consistent with tuberculosis (D.J. Robertson, unpublished). The availability of this recently excavated, well-documented, skeletal collection provided an excellent opportunity to evaluate mycolic acids as reliable biomarkers for ancient tuberculosis.

Adult mid-shaft rib fragments, with no lesions, and soil samples from both the surrounding area and the chest cavity of 21/210 (10%) skeletons were selected at random and examined for the presence of mycolates, using a method developed for the detection of tuberculosis in clinical material.[66] In essence, mycolic acids were extracted by alkaline hydrolysis and converted to anthrylmethyl esters; five of the twenty-one ribs tested were shown to contain components corresponding to mycolates on initial reverse-phase HPLC (rpHPLC). Subjecting the total mycolate fraction from these five ribs to normal-phase HPLC (npHPLC) showed the expected *M. tuberculosis* profile of α-, methoxy- and keto-mycolates. Analysis of the individual α-, methoxy and keto-mycolates from these ribs gave closely similar rpHPLC patterns, correlating well with those from modern *M. tuberculosis*. All soils were negative for mycolates. Mycolic acids, characteristic of *M. tuberculosis*, have been confirmed in 5/21 ribs under investigation (23.8%) and this correlates with the recorded frequency of 27.1% for the Newcastle Infirmary. These studies show that lipid biomarkers may allow the development of tuberculosis to be traced back into antiquity.[69]

The chromatographic profiles of the derivatives strongly suggest that the mycolic acids have survived with all their structural features intact. It will be of interest to investigate the detailed structural and stereochemical features of these acids by use of modern chemical and spectroscopic methods. There are preliminary indications that phthiocerol dimycocerosates may co-occur with mycolates in archaeological material (A.M. Gernaey, D.E. Minnikin and A.M.S. Ahmed, unpublished), opening up a whole new avenue for exploration. Similarly, analysis of mycolates and phthiocerol dimycocerosates may also be applied to the identification of leprosy in ancient bones.

6 Conclusions

Characteristic lipids are important structural components of mycobacteria, such as *M. tuberculosis*. The 60–90 carbon mycolic acids are a cornerstone of the cell envelope, interacting with other lipids to produce coherent hydrophobic defensive outer membrane resistant to common antibiotics. The properties of this unique organelle can be explored, using fluorescent probes which vary in their lipophilicity. The biosynthesis of mycolic acids is the target for the front-line drug isoniazid, and a number of other targets are being identified in these pathways. The extraordinary stability of mycobacterial lipids, such as the mycolic acids, is allowing the epidemiology of tuberculosis to be traced back into antiquity.

7 Acknowledgements

Studies on inhibitors of mycolic acid biosynthesis have been supported by Research Project Cooperative Agreement AI-38087 from the National Cooperative Drug Discovery Groups for the treatment of Opportunistic Infections, NIAID, National Instiutes of Health, USA through collaboration with Patrick J. Brennan and Gurdyal S. Besra (Colorado State University, Fort Collins, CO). Ultrastructural studies, using fluorescent probes, were funded by The Wellcome Trust in a project grant (GR 039680) and a prize studentship to N.J.G. Sensitive detection of mycobacterial lipid biomarkers was developed with grants from The Wellcome Trust and The Leverhulme Trust. Many studies have involved stimulating collaboration with Malin Ridell, University of Gothenburg, Sweden.

8 References

1. K. Duncan, *Chem. Ind. (London)*, 1997, 861.
2. J. Evans, *Chem. Br.*, 1998, **34**, 38.
3. C. C. Evans, in '*Clinical Tuberculosis,*' ed. P. D. O. Davies, Chapman and Hall, London, 1994, p. 1.
4. J. M. Grange, '*Mycobacterial Diseases,*' E. Arnold, London 1980, p. 7.
5. P. J. Brennan and H. Nikaido, *Annu. Rev. Biochem.*, 1995, **64**, 29.
6. T. R. Paul and T. J. Beveridge, *J. Bacteriol.*, 1992, **174**, 6508.
7. N. Rastogi, C. Frehel and H. L. David, *Int. J. Syst. Bacteriol.*, 1984, **34**, 193.
8. D. E. Minnikin, in '*The Biology of the Mycobacteria,*' eds. C. Ratledge and J. Stanford, Academic Press, London 1982, p.95.
9. M. Daffé and P. Draper, *Adv. Microb. Physiol.*, 1998, **39**, 131.
10. G. S. Besra and D. Chatterjee, in '*Tuberculosis; Pathogenesis, Protection and Control'*, ed. B. R. Bloom, American Society for Microbiology, Washington DC, 1994, p. 285.
11. P. J. Brennan and G. S. Besra, *Biochem. Soc. Trans.*, 1997, **25**, 188.
12. F. H. Stodola, A. Lesuk and R. J. Anderson, *J. Biol. Chem.* 1938, **126**, 505.
13. S. Hartmann and D. E. Minnikin, in '*Surfactants in Lipid Chemistry: Recent*

Synthetic, Physical and Biodegradative Studies', ed. J. H. P. Tyman, Royal Society of Chemistry, Cambridge, 1992, p. 135.
14. H. Christensen, N. J. Garton, R. W. Horobin, D. E. Minnikin and M. R. Barer, *Mol. Microbiol.*, 1999, **31**, 1561.
15. C. E. Barry, *Biochem. Pharmacol.*, 1997, **54**, 1165.
16. G. Middlebrook and M. L. Cohn, *Proc. Soc. Exp. Biol. Med.*, 1955, **88**, 568.
17. F. G. Winder, P. B. Collins and D. Whelan, *J. Gen. Microbiol.*, 1971, **66**, 379.
18. K. Takayama, *Ann. N. Y. Acad. Sci.*, 1974, **235**, 426.
19. K. Takayama, L. Wang and H. L. David, *Antimicrob. Agents Chemother.*, 1972, **2**, 29.
20. S. Kikuchi, T. Takeuchi, M. Yasui, T. Kusaka and P.E. Kolattukudy, *Agric. Biol. Chem.*, 1989, **53**, 1689.
21. H. A. Shoeb, B. U. Bowman Jr., A. C. Ottolenghi and A. J. Morela, *Antimicrob. Agents Chemother.*, 1985, **27**, 399.
22. K. Johnsson and P. G. Schultz, *J. Am. Chem. Soc.*, 1994, **116**, 7425.
23. K. Johnsson, D. S. King and P. G. Schultz, *J. Am. Chem. Soc.*, 1995, **117**, 5009.
24. J. A. Marcinkeviciene, R. S. Magliozzo and J. S. Blanchard, *Biochemistry*, 1995, **270**, 22290.
25. S. T. Cole, *Res. Microbiol.*, 1996, **147**, 48.
26. A. Telenti, P. Imboden, F. Marchesi, D. Lowrie, S.T. Cole, M. J. Colston, L. Matter, K. Schopfer and T. Bodmer, *Lancet*, 1993, **341**, 647.
27. M. E. Levin and G. F. Hatfull, *Mol. Microbiol.*, 1993, **8**, 277.
28. N. Honore and S.T. Cole, *Antimicrob. Agents Chemother.*, 1993, **37**, 414.
29. A. Scorpio and Y. Zhang, *Nat. Med.*, 1996, **2**, 662.
30. K. Takayama and J. O. Kilburn, *Antimicrob. Agents Chemother.*, 1989, **33**, 1493.
31. R. E. Lee, K. Mikušová, P.J. Brennan and G.S. Besra, *J. Am. Chem. Soc.*, 1995, **117**, 11829.
32. K. Takayama and N. Qureshi, in '*The Mycobacteria. A Sourcebook*', eds. G. B. Kubica and L. G. Wayne, Dekker, New York, 1984, p. 315.
33. S. Jackowski, J.E. Cronan Jr and C.O. Rock, in '*Biochemistry of lipids, lipoproteins and membranes,*' eds. D. E. Vance and J. Vance, Elsevier, Amsterdam, 1996, p. 43.
34. P. E. Kolattukudy, N. D. Fernandes, A. K. Azad, A. M. Fitzmaurice and D. Sirakova, *Mol. Microbiol.*, 1997, **24**, 263.
35. K. Magnuson, S. Jackowski, C. O. Rock and J. E. Cronan Jr., *Microbiol. Rev.*, 1993, **57**, 522.
36. T. Noto, S. Miyakawa, H. Oishi, H. Endo and H. Okazaki, *J. Antibiot.*, 1982, **35**, 401.
37. K. Bloch, *Methods. Enzymol.*, 1975, **35**, 84.
38. C. Ratledge, in '*The Biology of Mycobacteria,*' eds. C. Ratledge and J. Stanford, Academic Press, London, 1982, p.53.
39. S. Kikuchi, D. L. Rainwater and P. E. Kolattukudy, *Arch. Biochem. Biophys.*, 1992, **295**, 318.
40. P. R. Wheeler and C. Ratledge, in '*Tuberculosis: Pathogenesis, Protection and Control,*' ed. B. R. Bloom, American Society for Microbiology, Washington DC, 1994, p. 353.
41. P. R. Wheeler, G. S. Besra, D. E. Minnikin and C. Ratledge, *Biochim. Biophys. Acta*, 1993, **1167**, 182.
42. P. R. Wheeler, G. S. Besra, D. E. Minnikin and C. Ratledge, *Lett. Appl. Microbiol.*, 1993, **17**, 33.

43. H. Sasaki, H. Oishi, T. Hayashi, I. Matsuura, K. Ando and M. Sawada, *J. Antibiot.*, 1982, **19**, 397.
44. R. F. Waller, P. J. Keeling, R. G. K. Donald, B. Striepen, E. Handman, N. Lang-Unnasch, A. F. Cowman, G. S. Besra, D. S. Roos and G. I. McFadden, *Proc. Natl. Acad. Sci. USA*, 1998, **95**, 12352.
45. S. Jackowski, in '*Emerging Targets in Antibacterial and Antifungal Chemotherapy*,' eds. J. Sutcliffe and N. H. Georgopapadakou, Chapman and Hall, New York, 1992, p. 151.
46. S. Shimizu, K. Akimoto, Y. Shinmen, H. Kawashima, M. Sugano and H. Yamada, *Lipids*, 1991, **26**, 512.
47. A. L. Jones, J. E. Dancer and J. L. Harwood, *Biochem. Soc. Trans.*, 1994, **22**, 202S.
48. R. A. Slayden, R. E. Lee, J. W. Armour, A. M. Cooper, I. M. Orme, P. J. Brennan and G. S. Besra, *Antimicrob. Agents Chemother.*, 1996, **40**, 2813.
49. C. F. Huebner, J. L. Marsh, R. H. Mizzoni, R. P. Hull, D. C. Schroeder, H. A. Troxell and C. R. Scholtz, *J. Am. Chem. Soc.*, 1953, **75**, 2274.
50. N. P. Buu-Hoï, *Int. J. Leprosy*, 1954, **22**, 16.
51. N. P. Buu-Hoï and N. D. Xuong, *C. R. Hebd. Seances Acad. Sci.*, 1953, **237**, 498.
52. N. P. Buu-Hoï, N. D. Xuong and N. H. Nam, *J. Chem. Soc.*, 1955, 1573.
53. A. M. W. Hekking and K. K. De Voogd, *Antibiot. Chemother.*, 1970, 16, 128.
54. B. Phetsuksiri, A. Baulard, A. Cooper, D. E. Minnikin, J. D. Douglas, G. S. Besra and P.J. Brennan, *Antimicrob. Agents Chemother.*, 1999, **43**, 1042.
55. D. E. Snider, M. Raviglione and A. Kochi, in '*Tuberculosis: Pathogenesis, Protection and Control*', ed. B. R. Bloom, American Society for Microbiology, Washington DC, 1994, p. 3.
56. N. Bhatti, M. R. Law, J. K. Morris, R. Haliday and J. Moore-Gillon,. *Br. Med. J.*, 1995, **310**, 967.
57. R. P. O. Davies, K. Tocque, M. A. Bellis and P. D. O. Davies, Abstract from '*Proceedings of the International Congress on the Evolution and Palaeopathology of Tuberculosis*', eds. G. Pálfi, O. Dutour, and J. Deák, Tuberculosis Foundation, Szeged, Hungary, 1997.
58. R. Y. Keers, '*Pulmonary Tuberculosis: A Journey down the Centuries*', Ballière-Tindall, London, 1978, p. 1.
59. L. S Farer, L. M. Lowell and M. P. Meador, *Am. J. Epidemiol.*, 1979, **109**, 205.
60. R. T. Steinbock, '*Palaeopathological Diagnosis and Interpretation—Bone Diseases in Ancient Human Populations*', Thomas, Springfield, Illinois, 1976.
61. C. Roberts and K. Manchester, '*The Archaeology of Disease*', Sutton Publishing, Stroud, 1995.
62. R. A. McAdam, C. Giulhot and B. Gicquel, in '*Tuberculosis: Pathogenesis, Protection and Control*' ed. B. R. Bloom, American Society for Microbiology, Washington DC, 1994, p. 199.
63. G. M. Taylor, M. Crossey, J. Saldanha and T. Waldron, *J. Archaeol. Sci.*, 1996, **23**, 789.
64. W. L Salo, A. C. Aufderheide, J. Buikstra and T. A. Holcomb, *Proc. Natl. Acad. Sci. USA*, 1994, **91**, 2091.
65. A. G. Nerlich, C. J. Haas, A. Zink, U. Szeimies and H. G. Hagedon,. *Lancet*, 1997, **350**, 1404.
66. D. E. Minnikin, R. C. Bolton, S. Hartmann, G. S. Besra, P. A. Jenkins, A. I. Mallet, E. Wilkins, A. M. Lawson and M. Ridell, *Ann. Soc. Belg. Méd. Trop.*, 1993, **73** (Suppl.1), 13.

67. C. Roberts, D. Lucy and K. Manchester, *Am. J. Anthropol.* 1994, **95**, 169.
68. H. D. Donoghue, M. Spigelman, J. Zias, A. M. Gernaey-Child and D. E. Minnikin, *Lett. Appl. Microbiol.*, 1998, **27**, 265.
69. A. M. Gernaey, D. E. Minnikin, M. S. Copley, J. J. Power, A. M. S. Ahmed, R. A. Dixon, C. A. Roberts, D. J. Robertson, J. Nolan and A. Chamberlain. *Internet Archaeology*, 1998, **5,** http://intarch.ac.uk/journal/issue5/gernaey_index.hmtl.
70. A. Hardy, *Soc. Hist. Med.* 1994, **7**, 472.

Subject Index

Adiposity, 76
Adipostatins, 86
Adrenoceptor antagonists, β-blockers, 79, 91
  in slowing of plaque formation, 79
β_3 Agonists, 81
  synthesis of, 94
Alprostadil (PGE$_1$), 117
  biological properties of, 119
  competitor to Viagra, 124
Amino acids,
  empirical mixture as digestive inhibitor, 84
Anacardic acids, (13:0), (15:1), (17:1) and (17:2), 86
  inhibition of GPDH by, 86
  synthesis of, 98
Analytical instrumentation, 20
Anthocyanidins, 54
Antiaggregatory vasodilator, 119
Antiathersclerotic drugs, 77
Antituberculosis agents, 138, 145
Antimycobacterial agents, 139, 145
Antioxidant activity,
  effect of medium on, 60
  of quercitin re prevention of CHD, 61
Antioxidants, 50
  in lowering LDL cholesterol, 80
  primary, 50
  secondary, 51
Apigenin, 55
Appetite control, 81
  early work with amphetamines, 82
  role of anorectics, 82
  use of olestra, 82
Arachidonic acid, (AA), 10, 103, 109
  prostaglandins derived from, 116
Arbaprostil, PGE$_2$ analogue, 120
Atherosclerosis, 65
  experimental, 36
  influence in monkeys of *trans* fat on, 37
  influence in rabbits of *trans* fat on, 37
Attention deficit disorder, 102, 108, 109
Autoxidation, 47
  induction period in, 49
  mechanism of, 48

Bacillus, 130
Beef tallow, 6
Beraprost, PGI$_2$ analogue, stable prostacyclin, 119, 122
Bezafibrate, 79
  synthesis of, 91
Biosynthesis, 85, 116, 141
  fat, inhibition of, 85
  inhibition of fatty acid, 143
  of α-mycolic acids, 142
Biomarkers, 146
Brown adipose tissue, 81
Butter, 6

Calendula oil, 9
Camelina oil, 9
Carboprost, PGF$_2$ analogue, 121
Carcinogenesis, 35
  effect of *trans* fat on experimental, 35
  effect of *cis* and *trans* on mammary, 36
Catechin, 154

Index

Cerulenin, 143
Chalcones, 54
Chelation, 59
Cholecystokinin, 84
  synthesis of benzodiapine mimic, 98
Cholesterol, 38
  biosynthesis, 27
  inhibition of biosynthesis, 85
  lowering agents, 77
  plasma in man, 38
  serum, influence in man of *trans* fat on, 40
Cholestyramine resin, 79
Clofibrate, 79
  synthesis of, 90
Cloprostenol, PGF$_2$ analogue, 121
Cocoa butter, 6, 12
Coconut oil, 6
Colestipol, 78
(+)-Compactin, 78
  synthesis of, 88
Coriander oil, 9
Coronary heart disease, CHD,
  incidence of, 61
  HDL and decreased risk of, 65
  LDL and increased risk of, 65
  role of quercitin in prevention of, 61
Corn oil, 6
Cottonseed oil, 6
Crambe oil, 9
Cuphea species, 9
Cyclooxygenase (COX) enzymes,
  effect of iso forms of, 126
  role in PG production, 115

Dietary factor, docosahexaenoic acid for specific learning disorders?, 102
Dietary fatty acids, 65, 69, 71
  trans, influence on rat growth of, 34
Dietary intake, 72
Dihydroflavonols, 54
Dimorphotheca oil, 9
Dinoprost, (PGF$_{2\alpha}$), 117
  biological properties of, 119
Diseases, 76
  cardiovascular, 79
  role of prostaglandin levels in, 115
  tubercular, 130
Dimycocerosates, 136
Dinoprostone, (PGE$_2$), 117

  biological properties of, 119
Docosahexaenoic acid, DHA, 10, 102, 109, 112
  effect of high DHA fish oils on dark adaptation, 106
Docosapentaenoic acid, 109
Dyslexia, 102, 106, 107, 109
Dyspraxia, 102, 109, 110, 111

Eicosanoids, 116
Eicosapentaenoic acid, 10
Enprostil, PGE$_2$ analogue, 120
*Epi*sesamin, 143
Epidemiology, 42
  in study of *trans* fat, 42
Epoprostenol, (PGI$_2$), 117
  biological propeties of, 119
Essential fatty acids,
  biological role of, 103
  deficiency signs, 108
Ethambutol, 139
Ethionamide, 139
Euphorbia lathyris, 9
Euphorbia lagascae, 9
Extraction of oils, 11

Fat absorption,
  control of, 80
Fatty acid
  biosynthesis, 27
    inhibition of, 143
  composition of triacylglycerols, 6
  calendic, 9
  deficiency effects, 107
  dimorphecolic, 9
  erucic, 9
  essential,
    effect of *trans* acids on
    role of, 33
    signs of effect of possible deficiency of, 108
  lesquerolic, 9
  oleic, 9
  petroselinic, 9
  vernolic, 9
Fenofibrate, 79
  synthesis of, 90
(S)-(+)-Fenfluramine, 83
  synthesis of racemic and resolution, 95
  withdrawal due to side effects, 83

Fibric acid derivatives, 78
  for lowering plasma TAG and LDL cholesterol, 79
Fish oils,
  effect on dark adaptation, 106
  source of vitamins A and D, 107
Flavones, 54
Flavonoids,
  antioxidant activity of, 57
  dietary intake of, 56
  formation of radicals from, 59
  metal chelation by, 59
  pro-oxidant activity of, 61
  structure of, 53
Flavonols, 54
  antioxidant properties, 47
Fluoxetine, 83
  synthesis of, 96
Free radicals, 47
  classical route in autoxidation, 47
  in human disease processes, 52
  reactions, biological importance of, 52

Genes,
  role in excessive obesity, 85
Genetic modification, 9
Gemfibrozil, 79
  synthesis of, 90
Glycopeptidolipid, 136
Glycerol-3-phosphate dehydrogenase, (GPDH), 85
Groundnut oil, 6
GC-IRMS,
  current development in, 25
  GC-C-IRMS,
  applications, 29

Human immunodeficiency virus, 131
  role in resurgence of TB, 131
Hydrogenation, 8
  of vegetable oils, 32
  position of double bonds after, 33
Hypocretins, 85
  analogy with orexins, 85
Hydroxyphthioceranate, 136

Iloprost, PGI_2 analogue, 119, 121
Induction period, 49
Isotopes stable, 15
  terms and units, 18

Isoflavones, 54
Isomeric octadecenoates,
  distribution in tissues of rats, 34
Isoniazid, (isonicotinic hydrazide), 139
Isotopes,
  applications, 25
  costs of stable, 17
  hydrogen and carbon, 16
  measurement of, 17
  stable tracer, 15
Isotope ratio mass spectrometry, 22
  continuous flow-IRMS, 23
  dual inlet-IRMS, 22
  gas chromatography-combustion-IRMS,
  GC-C-IRMS, 24, 29
Isoxyl (ISO), 145

Kaempferol, 55
  oxidation of by superoxide, 60

Lard, 6
Latanoprost, PGF_2 analogue, 121
Lauric canola oil, 6
Learning disorders, 104
Leprosy, 131
Leptin,
  occurrence in fat cells, 85
  role in satiety, 85
Lesquerella oil, 9
Leucoanthocyanidins, 54
Leukotriene, 35
  antagonists and inhibitors of, 115
  lipoxygenase (LOX) derived, 115
Limaprost, PGE_1 analogue, 120
Linoleic acid,
  effect on serum cholesterol levels, 40
  oxidation of, 47
γ-Linolenic acid, GLA, 12, 104
Linseed oil, 6
Lipstatin, 81
  from *Streptomyces toxytricini*, 80
Lipaemia,
  postprandial, 65
  effect of dietary fatty acids on, 69, 70, 71, 72
  normal, 68
  exaggerated, 68
Lipid metabolism,
  constant infusion methods, 26

Index

enhancement in arterial wall, 80
study with stable isotopes, 25
Lipid peroxides, 48
Lipid oxidation,
  by haem compounds, 50
  metal-catalysed, 49
Lipids,
  composition of retinal, 102
  composition in human brain, 102
  cord factor, 136
  global resources, 1
  importance of mycobacterial, 130
  inhibition by GPDH of, 86
  lowering agents, 77
  phenolic, 86
  productivity, 3
  sulfatides, 136
Lipoprotein,
  chylomicrons, CM, 66
  high-density, HDL, 53, 66
  intermediate density, IDL, 66
  low-density, LDL, 66
    atherogenic potential of, 68
    inhibition of, 61
    oxidation of, 53
    quercitin in oxidation, 57
  triglyceride rich, TRL, 65
    effect of elevated levels of, 69
  very low-density, VLDL, 53, 66
Luteolin, 55

Meadowfoam oil, 9
Metabolism,
  study of lipid, 15
  of *cis* and *trans* fats, 33
  of essential fatty acids, EFA, 103
  of fat by thermogenic drugs, 81
  postprandial triglyceride, 66
  triglyceride, abnormality in, 73
Methods,
  analytical instrumentation, 20
  constant infusion, 26
  GC-MS, 20
Mevinolin, 77
  extraction of, 89
Misoprostol, PGE_1 analogue, 120
Mutton tallow, 6
Mycobacteria, 131
  cell wall, 133
  mycobacterial cell envelope, 132
  chemical components in, 138
  peptidoglycan, 132
Mycocerosic acid, 136
Mycolic acids, 134
  archaeological investigation of, 146, 148
  biosynthesis of, as a drug target, 141
  structures of representative, 134
Myricetin, 55

Nifedipine, 80
  synthesis of, 92
NuSun oil, 6

Obesity, 76
Oils and fats, 2
  annual consumption, 4
  bodily requirements, 4
Olestra, 82
Olive oil, 6
Orexins A and B, 85
Orlistat, 81
  inhibitor of pancreatic lipase by, 80
  synthesis of, 94

Palm kernel oil, 6
Palm oil, 6
Palm olein, 6
Pentadecylresorcinol, (15:0)-cardol,
  adipostatin A, 86
  synthesis of, 98
iso-Pentadecylresorcinol, adipostatin B, 86
  synthesis of, 98
Phenolphthiocerol A, 136
Phospholipids, 11
  role in membranes, 103
Phthioceranate, 136
Phthiocerols, A, B, 136
Phthiodiolone A, 136
Phthiotriol A, 136
Polyunsaturated fatty acids, PUFA, 71
  formation of prostaglandins from, 115
  importance in diet, 112
  in feeding supplementation, 104
Postprandial lipaemia, 65
Probucol, 80
  synthesis of, 93
Prostaglandins (PGs), 103, 115
  biological activity of, 115

Prostaglandins (PGs), *continued*
  chemical instability of, 118
  developments in the field of, 123
  drug development from natural sources, 117
  effect of small structural variations, 124
  medical and veterinary applications, 122, 127
  natural members, 119
  new areas of research in, 126
  novel drug delivery systems to overcome instability, 124
  novel formulations to overcome instability, 125
  PGD_2, 118
  rapid metabolism of, 118
  stable analogues, 119
  use in conjunction with non-steroidal anti-inflammatory drugs, 125
Pyrazinamide, 139

Quercitin, 55
  antioxidant activity of, 61
  glycosides, 56

Rape (high erucic), 6
Rape (low erucic), 6
Resolution, racemic fenfluramine, 95
Refining of oils, 8
Rifampicin, 139
Rice bran oil, 12
Rye, *cereale secale*, 86
  5-alkadienylresorcinols in, 86

Satiety drugs, 83, 84
Soybean oil, 6
Sertraline, 8
  synthesis of, 97
Serotonin, 83
Sesame oil, 12
Sesamin, 143
Sesamol, 143
Sesaminol, 143
Sesamolin, 143
Sibutramine, 83
  synthesis of 97
Stearidonic acid, 12
Sulprostone, PGE_2 analogue, 120
Sunflower oil, 6

Sunola oil, 6

Tetramycolated pentaarabinoside, 138
Thermogenic drugs, 81
Thiolactomycin, (TLM), 142
  inhibition of fatty acid biosynthesis, 144
Thiourea derivatives, as antitubercular agents in mice, 145
Tissue levels, 41
  of *trans* fatty acids in atherosclerosis, 41
Trehalose 6,6'-dimycolate, 137
Triacylglycerols,
  biological processes, 8
  blending, 8
  consumption, (disappearance), 4
  extraction, 8
  fatty acid composition, 6
  fractionation, 8
  from micro organisms, 10
  genetic modification, 9
  geographical sources, 3
  hydrogenation, 8
  inhibition of biosynthesis of, 85
  lauric, 6
  major oils, 1
  new seed oils, 9
  refining, 8
  speciality oils, 11
  world supplies, 3
Tuberculosis, 130
  ancient, 146

trans Unsaturated fat,
  effect on human plasma cholesterol levels, 38
  in American diet and mortality, 43
  in health and disease, 32
  role in cholesterolemia, 39
  vaccenic acid, 32
Unoprostone, PGF_2 analogue, 121

Verapamil, 80
  as calcium channel blocker, 79
  synthesis of (2S)-(−)-, 92
Vernonia galamensis, 9

Xenical, 80